# LEADERSHIP AND TRAINING FOR THE FIGHT

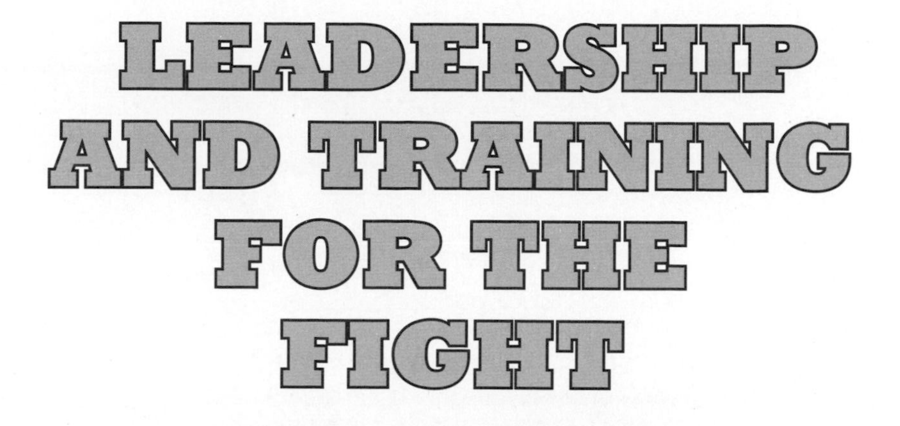


## USING SPECIAL OPERATIONS PRINCIPLES TO SUCCEED IN LAW ENFORCEMENT, BUISINESS, AND WAR

**MSG PAUL R. HOWE**
**U.S. ARMY, RETIRED**

Skyhorse Publishing

Skyhorse Publishing books may be purchased in bulk at special discounts for sales promotion, corporate gifts, fund-raising, or educational purposes. Special editions can also be created to specifications. For details, contact the Special Sales Department, Skyhorse Publishing, 307 West 36th Street, 11th Floor, New York, NY 10018 or info@skyhorsepublishing.com.



www.skyhorsepublishing.com

20 19 18 17 16 15 14 13 12 11

Library of Congress Cataloging-in-Publication Data

Howe, Paul R.

Leadership and training for the fight : a few thoughts on leadership and training from a former special operations soldier / by Paul R. Howe.

p. cm.

ISBN 978-1-61608-304-5 (alk. paper)

1. Leadership--Handbooks, manuals, etc. 2. Command of troops--Handbooks, manuals, etc. 3. Tactics--Handbooks, manuals, etc. I. Title.

UB210.H68 2011

658.4'092--dc22

2011008333

Printed in China

I would like to dedicate this book to the following:

**MY WIFE** has helped me carry my rucksack through life with unwavering support and dedication. Without her consummate stability, I could not have made this journey with the success I have found.

**THE MEN AND WOMEN WHO SERVE IN OUR ARMED FORCES**. The army provided me a home where I could learn and grow for a great part of my life. It pushed me to improve and Special Ops gave me a home to grow. Special thanks go to the noncommissioned officers (NCOs) who set the bar and taught me to push farther and faster with greater precision, not because it was required, but because it was the right thing to do. I am also grateful to them for affirming the importance of doing what is right versus what is popular.

**THE LAW ENFORCEMENT COMMUNITY** is the glue that holds our country together. Generally underpaid, undertrained, and overworked, they continue to protect and serve for the right reasons.

# CONTENTS

# NOTE TO READER

**OUT OF COURTESY** to my former unit and their sensitive nature, many of my descriptions are referred to as Special Operations. I have attempted to sanitize this work as much as possible. Having said this, all the incidents I describe here are as accurate as I can recall. Also, remember that my perception is based on my cumulative experiences in the military and in the law enforcement community. My view may vary from that of others, and my expectations of an individual or a unit may weigh in on this. A former commander often said, "To whom much is given, much is expected." I have grown older and wiser over the years, gaining a bit more insight into and understanding of combat leadership. With that said, I try not to judge soldiers on an individual action; rather, take a comprehensive look at their training, leadership, and their individual hearts.

In addition, I have also laid out my opinions for the best way to structure high-risk tactical courses in this book. I have developed these courses throughout my years of instruction with combat shooting and tactics (CSAT). While I describe various techniques, I will limit any description of tactics as the bad guys have a tendency to study our work. Sometimes, I will be generic, and it will be so on purpose.

I will also describe some near misses and training accidents in this work. This is not to demean or cast a shadow of doubt over those involved. It is necessary to learn from our mistakes. If we cover them up, they will be repeated at the same price — the loss of a human life. My goal is to always provide safer and more efficient training techniques to fix the problem. These safety techniques will come through a logical stair-stepped methodology.

# PREFACE

**AFTER GIVING IT** much thought (and with a little practical experience), I have come to the conclusion that our society or way of life will not come to an end because of a natural disaster or at the hands of a superior enemy, but rather through a lack of leadership and initiative on our part. Leadership is what has made our country great, and what will break our country if enacted poorly.

Leadership permeates all folds of military, law enforcement, and tactical training operations. Generally, success or failure of the mission can be tracked down to either a leadership genius or void in the selection or training of the personnel at the individual, team, or organizational level of the tactical element. Leadership is not rocket science. Positive leadership requires only that the leader follow simple and common-sense rules and practices to ensure the success of an assigned mission.

I have been prodded over the years by friends, students, and colleagues to write a book on my thoughts and experiences. I hesitated initially for security reasons, but I also felt that Mark Bowden did an exceptional job of tying together all the individual and leadership actions in his book *Black Hawk Down*. These were actions that military leaders at the time attempted to sweep under the carpet to avoid the political fallout. Also, I hesitated because I had doubts in my writing ability; I wasn't confident I could put together my material in a comprehensive and easy-to-understand package.

I eventually gave in and decided to embark on this endeavor head-on. More importantly, I believe I now have enough experience and understanding to write in a clear and focused manner. Throughout the writing process I have kept in mind the words of an old friend, former Secret Service agent Carl Kovalchik: "I am not trying to learn new stuff; I am just trying to remember what I have."

With that, I want to try and capture the focus and aggression I had as a younger man and not try to reinterpret it as that of a "kinder-gentler" middle-aged ex–action guy. I knew that when I started this project it would be a work in progress, as is life. I will reference sources in this work that I deem accurate and viable and inspired and helped me in my learning process and survival. I will use a great deal of my life experiences that I can recall and, where relevant, discuss my mind-set and how it helped me survive. I have found that your mind-set plays a key point in action, interaction, understanding, and development. As I get older, I find that I have gained a clarity of thought and the patience to put my thoughts down on paper. By writing this book, I have been able to recapture that which escaped me as a younger man, when testosterone and excess energy tended to get in the way of clear communication.

This book tackles leadership and training on two fronts. Part I discusses how to identify, build, and utilize the skills that will allow individuals and teams to efficiently and effectively achieve their goal. Part II focuses on training methods and techniques that will allow those same individuals and teams to practice effective combat and leadership skills in the field.

Why write on leadership? Too many times in our lives, we encounter ticket-punching ladder climbers and those who seek the role of leader for the wrong reasons. Whether it's politics or the military, many times we find the system skewed by leaders who wish to guide men but have cheated in their own lives and experiences. Many are self-serving cowards who climbed over good people to attain their goals. Poor leadership in the business world will cost you time and money. In the military and law enforcement world, it will cost the lives of our mothers, fathers, sons, and daughters.

This book is based not on leadership theory but on a practiced system used and tested under the worst-case scenario—combat. In my eyes, the validation of a leadership system culminated on October 3–4, 1993, in a one-sided battle in Mogadishu, Somalia. There I saw leadership at the individual, team, and organizational levels both excel and fail. Within these pages I hope to enlighten you as to the techniques that succeeded and those that fell short, and why. I will do my best to show you the path to follow to achieve success in personal combat and on the battlefield as a unit. There are many more battles to be fought and won to ensure our survival and way of life.

Why write on training? Training, like selection, is a never-ending process. We begin learning on day one of our life. We spend much of our time attempting to streamline, simplify, and modify our personal system to make it more efficient and stress free. Books and classes abound on how to learn at the individual level and how to manage time more efficiently. The next step in the learning curve is to teach. I have witnessed both good and bad teachers. Too many times teachers, or "instructors," want to wear the title or the T-shirt, if you will, to satisfy their own egos. Some do it for the paycheck. Some do it because it is a natural progression in their jobs. Others do it for the passion and as a way to give back. The latter are the people I am writing this book for.

My intent in writing the second part of this book, or manual, is to help out trainers in several areas. First, I want to lay down a simple and streamlined guide on how to be a good student, how to learn, and how to organize what you as a student have absorbed into an efficient package. I want to teach instructors how to create efficient and logical instructional modules and conduct classes where a student can gain the most in the least amount of time. I believe that instructors can always refine their instruction techniques. In this work, I will address both classroom and field training exercises. I will elaborate on the key points that have helped me become successful in my profession.

In today's world of budget cuts and time constraints, it is critical that instruction be clear, concise, logical, and safety conscious. The military, police departments, and corporations all lose valuable training time and dollars to poor trainers and weak programs. A student must unlearn what was taught in a poor training program, and then devote additional training time and funds to get things right.

My profession is that of instructing in high-risk training, where failure to teach properly or to stair-step the material can have deadly consequences. Many a student has been injured, crippled, or killed during high-risk training, and it's the instructor who takes the blame. I wish to reduce those numbers and get students safely to the streets or to the battlefields, so they can better serve our country and then return home to help others.

Again, the information presented in this book is not based on training theory, but on a practiced system used and tested in combat. The validation of these training systems comes from thousands of hours of high-risk training scenarios that I have been fortunate to have participated in both as a student and as an instructor. We need to safely and efficiently produce better soldiers and law enforcement officers who dominate in their battlefields and return safely home.

# INTRODUCTION

## ACCELERATING THE LOOP

*The speed must come from a deep intuitive understanding of one's relation to the rapidly changing environment.*
—Robert Coram, *Boyd: The Fighter Pilot Who Changed the Art of War*

- **TWO COMBAT SCENARIOS — EFFECTIVE AND INEFFECTIVE TECHNIQUES**
- **BOYD'S THEORY AND INDIVIDUAL COMBAT**
- **INSIDE YOUR ENEMY'S MIND**
- **OBSERVE**
- **ORIENT**
- **DECIDE**
- **ACT**
- **CONCLUSION**

## TWO COMBAT SCENARIOS — EFFECTIVE AND INEFFECTIVE TECHNIQUES

Colonel John Boyd has been described as one of the principal military geniuses of the twentieth century, and yet few know his name or accomplishments. Nicknamed "Forty Second" Boyd, because it was his standing bet with pilots that from a position of disadvantage, he would turn the dogfight around and be on his opponent's tail within forty seconds. The rumor is that he never lost. More important than the challenge was Boyd's ability to articulate his winning strategy and teach it to others.

Boyd's OODA Loop (observe, orient, decide, act) has been explained in several different fashions and interpreted by each audience in a different way. On paper the loop takes a more linear approach consisting of observe > orient > decide > act. A simplistic view of the cycle, or loop, would be that you *observe* the enemy and their actions. Next you would *orient* yourself or your forces into a more favorable position. You would then *decide* on your course of action. Finally, you would *act*.

According to Robert Coram, the author of *Boyd: The Fighter Pilot Who Changed the Art of War*, Boyd's intent was not a speedy mechanical solution, but rather to implement the cycle in a way that you could get inside your opponent's mind and to disrupt their decision-making cycle by ensuring that they are dealing with outdated or irrelevant information that does not allow them to make a decision or to function.

Since its creation, Boyd's OODA theory has been proven effective in one combat situation after another. While the methodology for applying his technique changes with the times, the brilliance of the theory lies in the fact that the core principles can be effectively applied in almost any combat situation. That is why I have placed Boyd's Loop at the center of my mental and tactical training for over twenty years.

Below are two different scenarios—one with a positive outcome and one with a negative outcome. As you read them, keep in mind that this book is intended to help leaders develop both mental and tactical skills that will allow for the most effective combat force possible. Observing real scenarios makes it possible to understand the leadership skills Boyd's system fosters; it's also possible to see how tactical training fails when OODA is not applied. Boyd's Loop is the most essential tool for anyone looking to develop their leadership skills or hoping to teach tactical training to others.

---

## SCENARIO 1: NEAR MISSES

It was the dead of night, and we were flying low, trying to avoid TV and radio antennas that could be a showstopper should we collide with one. I was sitting to the rear of the pod, and I got the one-minute index finger signal from the pilot. I started looking ahead at the buildings, trying to pinpoint our target. It was a walled compound that held a potential hiding place for our top personality. Flying a UN flag, this target was reputed to be dirty.

In the same compound was an adjacent multistory building that was a secondary target and potential hiding place of our target. Teams were assigned to enter and clear this building as well as our primary target building. I was still trying to pinpoint our building, which had a distinctive roof design and a certain type of antenna. I was scanning hard on our final approach, and things did not look right. Instead of depositing our entire force in one building, the pilots let us off in three separate buildings. I saw a bird to our rear let off a team on a building forty to fifty meters behind us, inside another walled compound. I knew this could be dangerous because our chain of command had briefed us that security forces were present on buildings either side of our target.

I jumped off my pod, hitting the ground and getting on a knee. We had a three-foot parapet at the top of our building, which protected us from hostile fire. At the base of the inner parapet were floodlights that shone out through the slits in the parapet wall. I tried taking the light out with a butt stroke from my weapon stock, but the butt of my "plastic" stock just bounced off. I found a piece of broken concrete within arm's reach and gave it a good smash, and the light went out.

I started scanning to our rear and saw that the team behind us was on top of their building and appeared to be shooting out their floodlights. I could see the backlit silhouette of an operator. The team was indeed on top of the roof, shooting out their lights with their CAR-15's. All of a sudden, someone fired two shots next to my head, the tracers screaming past the team member on top of the building to our rear. I struck the shooter (my partner) in the helmet with a backhanded closed fist and shouted "Cease fire! Those are friendlies!" Evidently, he had not been scanning deep enough with his night vision goggles (NVGs) and had not seen the bird let one of our teams on the roof to our rear.

The two rounds fired were tracers, and I could not see if they connected with one of the team members because as they streaked by, he dropped down so fast it looked like he was hit. The rounds kept going and I could see an inbound Helo jerk hard to avoid the rounds. My heart sank at the thought of a friendly-fire casualty. I could understand why my partner shot; he heard and saw gunfire, saw the silhouette with a gun, and was briefed that the building was hostile. I was torn. One part of me hoped that he had missed, but another part recognized that his poor aim created a different dilemma for me as a team leader (TL).

I knew that we were probably on the wrong building, but did not want to stay on top and become a bullet magnet or get pinned down from the ground. I led the team down a rickety spiral metal staircase outside the building. The team started taking names and kicking ass. People had been sleeping in the courtyard and were coming out of the woodwork, waking up and squinting at us. I ran into one guy and he spun off me. I turned around and flashed him with my gun light. When the light hit him, he acted as if I shot him. He put his hands up in front of his face, and he crumbled backward to the ground.

Team members were starting to flex-tie individuals, and I was trying to link up with another TL to find out where the target building was in relation to our positions. I found another TL from another building and asked him if any of his guys were hit; I was still worried about the two tracer rounds. He said his guys were good to go.

Just then, our bird reappeared and hovered above our roof.

Outstanding!

The pilots had come back to get us and reposition us on the right roof. I gathered the team, and we climbed the shaky staircase. I could only hope that it would support our weight one more time. We got to the roof, got on the bird, and took off for the right building. A couple of minutes later, we were at our target, and I could see several teams had already arrived.

To my amazement, they had not blown off the roof door or penetrated the building. It seemed the assault force was spread over several buildings, and everyone got there late. The door was blown off and we started clearing. We cleared the third and the second floors and linked up with a team working their way up from the bottom. We went back to the roof and found an exterior second floor door that was not open, and we threw a fast rope over the edge and had one guy body-wrap it while we slid down to the next landing. We just wanted to ensure that we did not miss anything.

Upon securing the building, we gathered up our package and cleared the courtyard buildings and vehicles, which was our secondary mission. Afterward we conducted a link-up with our perimeter teams at the front gate. We consolidated our assault force and moved about a block on foot to a hasty helicopter-landing zone in a walled courtyard.

Upon arrival, I started asking questions as to whether or not the area had been cleared. The problem was that building adjacent to the landing area had ten to fifteen rooms that were all facing the bird, and this building had not yet been cleared. The team and I started clearing the rooms to ensure our incoming bird would be safe. Just as I thought, there were families in some of these rooms that were frightened and huddled in corners. We conducted a hasty search for weapons and then motioned for them to stay put. The last thing you want is a gunfight in close quarters with a helicopter in the middle of it with guns pointing in all directions or civilians running into the action or into a tail rotor.

## AFTER-ACTION COMMENTS

### *SUSTAIN*

- Aggressiveness.
- Follow through on mission.

### *IMPROVE*

- Friendly-fire briefings and rules of engagement (ROE).
- Look harder and be sure of your target.
- Communicate faster to team members wearing NVGs as they have tunnel vision, and you both might be seeing different things.
- Counsel subordinate on shooting and missing shots.

## SCENARIO 2: DEATH IN THE LOOP

I have watched the video countless times and have now watched the faces of my students watching the video. The video shows a deputy sheriff in the South who is conducting a traffic stop on what appears to be an irate and possibly drunk older man with a beard. The car camera recorded the video. The sound is possibly as disturbing as the images.

The deputy begins his verbal escalation in an effort to control the situation, and he becomes stuck in what I call a verbal loop. The deputy's mind has seized up, and the subject is not complying with his instructions. He continues to shout instructions to the subject, raising his voice almost to a screech. He does not hear what the subject is saying. The subject is cursing and says he is a Vietnam vet. This should be sending signals to the officer to request for backup and to either get hands-on or break contact and retreat to a defensive position.

The subject walks to the driver's side of his truck and takes out a rifle. The deputy is aware of this and commands the man to stop, but the man doesn't listen to him. The subject gets into an offensive crouch position with his rifle. The deputy begins to fire; he misses, and continues to miss. The subject pushes the fight, moves toward the deputy, and engages him from the front of the squad car. You can hear the deputy scream as he is hit.

The subject then moves between the right front of the patrol car and his truck and is either checking or reloading. The deputy shoots again at the subject and misses. The subject approaches the deputy and fires several more shots. The deputy screams. The subject leans over the front of the patrol car and executes the deputy, and then moves back to his vehicle and drives off.

## AFTER-ACTION COMMENTS

*SUSTAIN*

- None noted.

*IMPROVE*

- Being caught in verbal loop.
- Physical escalation.
- De-escalation skills.
- Chemical escalation.
- Marksmanship.
- Decisiveness and aggressiveness.

---

Personally, I have watched too many situations just like Scenario 2 unfold on countless video clips that come from cameras mounted inside patrol cars. They catch the good and the bad of what happens, and many times the audio is caught along with the video. In short, I hate to see the good guys die or get hurt and the bad guys get away unscathed. It angers me to no end. Instead of complaining about it, I intend to show you how to better train and prepare future officers.

The deputy sheriff described in the above scenario was caught in Boyd's Loop. He observed and oriented. He was hesitant to decide. When he finally did act, his technical skills were not where they should have been for such a high-stress situation. I see two immediate areas lacking in this scenario. The first being his decision-making process, and the second his technical skills, namely his proficiency with his weapon. Deficiencies in these two areas cost him his life.

As to why, it is very simple. I have been fortunate to be drawn into a profession and career field that I am passionate about. I believe in the cause and those who volunteer to serve.

Many times we fail them with our instruction. With that point, I have seen too many die because of poor classroom instruction, ineffective training exercises, and weak or minimal sustainment training.

With this book, I hope to provide the answers to the test in both classroom and field training. We have a duty and responsibility to train those to come after us, not only in terms of technical skills, but also in teaching and leadership skills.

Students generally begin learning technical skills as a means to solve a problem. Once students have learned and mastered individual skills, they may be tasked to start teaching these skills to others. Once the new instructor has a solid grasp of these skills, he may begin to train other instructors.

For readers looking to foster both leadership skills and tactical proficiency, the path I am going to outline for you is harder and tougher than others. It is one that no one will thank you for. In the end, it will help ensure that you go home each day to the spouse and fussing kids. It will help you to grow old and know you did the right things for the right reasons.

## BOYD'S THEORY AND INDIVIDUAL COMBAT

Boyd's successful use of the loop involved compressing time (the loop) in a faster manner than your opponent. He said, "The speed must come from a deep intuitive understanding of one's relationship to the rapidly changing environment." In effect, the one who learns to move through the loop faster than their opponent will generally win.

There are some circumstances where this may not apply, but generally it will ensure your success. A case cited by Coram is that of General Patton and WWII tactics and feats. Instead of giving specific mission orders to seize a specific spot of ground, Patton suggested that a broader mission statement of "intent" be used. This intent would enable forces on the ground to use their own initiative and exploit the enemy's weaknesses as they saw fit. This empowerment, aimed at the lower levels of leadership, is extremely successful. The use of this technique also makes for more efficient strides and gains in battle.

Instead of throwing all assets or forces at a particular strongpoint, Patton pushed for his soldiers to receive an element of empowerment and trust. He allowed them to use their heads and select the path that would enable them to accomplish their mission with the greatest chance of survival. What a concept! This in itself would create a force for planning, initiative, and aggressiveness, which would have a direct impact on survival. In my eyes, such a force would bring out critically needed motivation and ideas into most combat scenarios.

In Patton's case, he would attack through the initial layers of the enemies' defense, ignoring his flanks, creating chaos not only there but throughout the enemies' forces, both with the troops on the line and with the support troops. It is the above case that is in the minds of the support troops. They are greatly affected by the chaos and the fear of the unknown, whom we can mentally catch and exploit in the OODA loop.

## INSIDE YOUR ENEMY'S MIND

Coram quotes Boyd as saying, "Machines don't fight wars, terrain doesn't fight wars. You must get into the minds of humans. That's where the battles are won." This may seem a bit deep, but it is actually pretty simple. Once you learn to use the loop, you can employ simple and effective tactics that will shock and disorient your opponent long enough for you to seize the initiative, better position yourselves for the act phase, eliminate (kill) them from the loop, and then find another victim to apply your loop to. This concept applies to the *strategic level*, the *battlefield*, the *team level*, and the *individual level* of combat.

To emphasize the practicality of Boyd's Loop, imagine the following scenario: An individual is "running and gunning" on an assault, and your force is tasked with the goal of pushing through an enemy fortification. In this case, the enemy is on the defense, and they have prepared a suitable response plan to repel any attack. Their first line of defense is to fire mortars or artillery at the attacking force. The next line of defense is for individual soldiers to directly fire their rifles and rocket, propelled grenades (RPGs) at the attackers. Once the attackers get into hand-grenade range (thirty-five meters), they will throw these killer eggs in your direction, all while still firing their primary weapons. Once you get too close, the enemy will fall back to alternate positions and continue to engage you.

The enemy intends to do all this in an orderly process, so this is where we bring chaos and disorder to their plan. First, we drop cluster bombs on their positions. Next we call in some artillery shells, some that air-burst and some that go deep in the ground and then explode, creating large craters, moving tons of earth in the span of a second. We then fire white phosphorus rounds, which burns everything that they come into contact with, to prevent the enemy from observing our approach. By this time, all but the most hardened enemy soldiers have turned into frightened animals. Their artillery is neutralized. The landscape does not look the same as it did when they went into their holes. They now have limited visibility with the area on fire and smoke blanketing everything. They cannot see the attackers, but are taking effective fire from them. A bleak picture indeed.

Now picture this: The entire attacking force has bypassed you, thinking you are not significant; and they are pushing deep into your ranks. In effect, you are surrounded. Think about your mind-set now. Dead and wounded scattered in the rubble strewn landscape. Occasional mortar rounds are still exploding, letting you know that someone is still thinking of you and they still have you under their guns. More armored vehicles passing by, beyond the range of your guns. A majority of regular soldiers would probably think about giving in.

At the individual level, it gets even more intense. You hear the sound of explosions getting closer. As you poke your head around a corner to look, the concrete explodes in your eyes where the bullets strike the wall, narrowly missing your head. The last glimpse you see as you fall backward a step and try to shake it off is of several shapes closing in on you. You learned your lesson about sticking your head out around corners. You regain your composure and prepare to angle your weapon around the corner and let loose a burst of automatic fire to give the attackers something to think about.

*Yeah*, you think to yourself, *I'm doing good*. Just then, you notice a small round and green object roll past the corner about three feet from you. Your eyes focus on it, and you realize it is a fragmentation grenade. And then it explodes. You feel the steel pushing through parts of your body and you fall backward from the blast. You try to stand, but for some reason, your upper leg seems to push past your lower leg and touch the ground. You try to take a breath, and you realize that it is like trying to breathe through a sopping-wet handkerchief. You then realize a great deal of your lungs are sticking to the wall behind you. As you gurgle for air and as your vision dims, the last picture your mind snaps is that of an American flash suppressor spitting fire in your face. Too late, someone has moved through the loop faster and more efficiently than you.

For a domestic officer, the details of combat may be different, but the application remains the same. While performing routine patrol duties, an officer gets the call of an active shooter in an office building. The officer acknowledges the call and begins to think of several things. First, route selection is critical. If you cannot get to the fight, you cannot fight. The officer must know the outside area of the target building and must determine a route that will get him there safely and put him in a tactically superior position. He must calculate traffic for that time of the day and determine the best way to approach so he can get as close to the target building as possible, using cover and concealment. He should be looking at how to make himself a hard target should the shooter emerge on his approach or shoot at him from a window or a door of the building.

So how does the soldier end up on the right end of the flash suppressor? How does the officer avoid becoming another casualty of domestic warfare? *They learn to move through the loop faster, more efficiently, and with more focus than the opponent.* How can this be accomplished? *By breaking down the entire process into simple and decisive actions.*

Will technology give us an edge? You bet. At times technology will aid in all portions of the loop. Within the next few sections, I will share my view on how to best accelerate the loop at the individual, team, and organizational level. I will also give you a few simple techniques that will help you to accelerate your personal, team, and organizational journey through the loop.

## OBSERVE

### *Individual Level*

The first step in accelerating the loop at the individual level is to teach the person to observe, or become conscious of their surroundings. I will use urban environment as an example. Subconsciously, after years of training and combat, I learned where to look so as not to waste valuable time scanning. the entire area. Instead of looking at everything in the environment—walls, trees, cars, roofs, etc—you focus on where a person could shoot at you from.

If you try to look at everything, *you are already getting behind in the loop* because you are overloading your brain with useless information and images. It is not from a blank wall or on the ground in front of you that an enemy gunman is going to shoot from, but rather from that doorway on the right, the wall to your center, or the window to the left. Knowing this, you can focus on the battlefield and narrow your scanning to relevant information. You might have caught movement or fire from a window and suppressed it. The reason you catch movement in the window is that it has a higher priority in your scanning sequence than other areas and you may scan it twice as much as, you would the minor-threat areas.

### *Team Level*

The same process can be used for team movement, but your point or lead person is focused on the front, while other members are going to be watching their sectors (assigned areas), such as an alley to the left or right. These areas might be constantly changing and shifting. When looking at the alley, team members should be concentrating on the edges of buildings for signs of people. Muzzle flashes, dust clouds, and shadows are indicators of threats. By looking at the edges and not the center of the alley, team members pick up threats faster than if they scan from the blank center of a street to a wall.

### *Organizational Level*

Simply by expanding the above philosophy that I have described, your organizational element will reap the benefits of individual and team observations. Your empowered elements will see farther and faster than their adversaries and report this information back or act on it, depending on what you instructed them to do in their mission orders.

---

*Training for Observation*

Today's technology and optics can greatly enhance this portion of the loop and can cut your visual pickup time for threats and/or targets. At the individual level, a variable scope on a soldier's rifle in an urban or open environment may aid the speed of observation. At a distance, I recommend that optics be run on the lowest power because an immediate threat may present itself and take up the entire field of view. Use your normal scanning and vision to alert you to danger areas and then use the camera on your scope to increase the magnification or view so you can more easily see your area or point in question.

I remember one action when the bad guys knew our ROE. They knew that if we could not see a weapon, we could not shoot. They tried using it against us by sawing off their AK-47s (Kalashnikov rifle) and wrapping them in a sheet or towel. It's almost impossible for an unaided eye to see a weapon one or two blocks away. However, with a small bit of magnification, it became easy to see the outline of a front sight post, the pistol grip, or an entire weapon. This visual allowed us to comply with ROE and service those threats. I returned from this trip and promptly bought a 1.5 x 4.5 compact scope for just this reason. The same theory goes for our long-range and standoff weapons, including tanks and helicopters. The farther out we can see, the farther out we can engage. This will generally speed us through the loop and ultimately impact on the cumulative tactical time.

How can we improve our observation and scanning sequence at the individual level? Simple—develop scanning exercises that teach individuals to see fast. I routinely ask students whether they shoot first or see first. The answer is obvious. They must see first before they can shoot. In one portion of the scanning sequence, I train students to discriminate.

In the old days when I was first assigned to Special Ops, they taught us to look at the hands first. As emphasis was placed on our shooting skills and speed, our shooting developed faster than our scanning skills. As a result, discrimination suffered and the number of friendly-fire accidents increased. Individuals were looking at hands, bringing their weapons to center mass of their target, squeezing one to two rounds off, and then their brain would catch up and say, "Uh-oh, that is a friendly officer." I have witnessed this with paintball, simulations, live-fire exercises, and in combat. Consequently, I have changed my scanning process to look at the whole person first, see who it is and then collapse to the hands.

The drill I use to enhance an officer's scanning process is a simple one. I have fifty premade targets that I set up in a classroom-size room. Twenty-five are set up in depth on one side, and twenty-five are on the other. On the command to enter, two students enter and go to their sector and must scan the twenty-five targets for the one with the gun. When all is said and done, each target must be scanned three times each. First, "whole person," and then you collapse to right/left or left/right hand.

With that simple drill on one side only, seventy-five scans need to be made. I record a student's first attempts at the drill, their entry and scanning sequence. Usually the same problems are present: weapon too high in the visual plane, scanning too fast, not scanning far enough, etc. I teach the students to develop a systematic scanning sequence that goes from hard points in the room (points that don't change from room to room), and push them to do the whole person and hands. As they hit the room several more times, they practice their system, and their speed increases.

Another problem that comes into play is the visual overload upon entering the room. I do not allow my students to see the room prior to entry. Upon entry, they are visually overwhelmed and do not resort to the scanning system that I teach in class. Routinely they will be standing right next to the threat and not even see it, even after they give an "all clear." I intentionally attempt to visually overwhelm them and force them to consciously focus their scanning on pertinent information that you would find on the battlefield or in a raid-type scenario.

This drill has humbled many a seasoned officer. In addition, it is inexpensive to set up and can be practiced without firing a shot. This same drill can be performed with a team and taken to the next level by including video graphics. As in real life, those practicing most video games know where the threat is going to expose itself from, and we in turn set ourselves up in the most advantageous spot to engage it. This is where the *orient* step comes into play, allowing students to see and place themselves in a tactically advantageous position.

---

## ORIENT

### *Individual Level*

The individual orient phase can be thought of as posturing, or preemptively assuming a tactically superior position. In the observation phase, you have already observed and are aware of the threat areas. You now need to adjust your individual position and maneuver to an offensively superior position. Note I did not say defensive; I said offensive. You must position yourself in such a way as to maintain maximum cover and aggressively outmaneuver the threat.

Take for example the street movement described in the earlier section. Individuals should move in such a way as to visually obscure their movement from the enemy while keeping the enemy in constant view. Gunfights are nothing more than the successful use of geometric angles to protect yourself as much of the time as possible.

During movement, try and expose as little of your body as possible, for the shortest time possible. Make yourself a small/hard target. In other words, make the bad guy expose more of himself to you. This increases your chances of hitting him sooner and faster. Use natural and man-made objects as cover to make these movements. Put trees, walls, cars, etc., between you and your objective in order to get closer, thereby denying the enemy the physical and mental advantage.

### *Team Level*

At the team level, do the same, but lead the team on the most covered and concealed route to the enemy. For an added advantage, your team should have a decisive reaction drill to contact and automatically deploy when taking fire. This should be a type of nonverbal communication. For example, if team members see or hear you shoot, that is a nonverbal form of communication, and they know to automatically execute the planned drill. We thereby eliminate the verbal command from the process, further expediting our ride through the loop. This can take the form of an immediate action drill (IAD) such as being fired upon when moving to the front door of your planned target. We plan on this because we have worked in this environment and we know that once the first round is fired, you can't hear (auditory exclusion), and the visual becomes your primary source of information.

### *Organizational Level*

The organization is an extension of the team, and all the same rules apply. Individual teams making up the organization should be applying all the same tactics and techniques in their individual battles or firefights to ensure their desired outcome.

---

### *Training for Orientation*

Training at the individual level is relatively simple, reinforcing all the basic skills dealing with the use of cover and concealment. Both the soldier and the officer should understand cover and concealment and when to use them. *Cover* is an object—man-made or natural—that will protect the soldier from fire, such as a concrete wall or a dirt berm. Cover will stop bullets and shrapnel. *Concealment*, on the other hand, is simply a visual barrier that keeps an opponent from seeing you, but it will not stop bullets. Concealment might consist of hedges or shrubs around a house. It is critical to ensure that the individual understands what cover is and how to use it. He should constantly put it between the enemy and himself to greatly enhance his chances of survival.

Training the team for this phase is a bit more complicated. The team must learn the safety aspects of fire and maneuver, weapon handling, and when to shoot, etc. Team members must continually scan for cover, both individually and as a team. A team who decides to hide behind one tree makes it easy for the enemy to put effective fire on them, especially when they are in a line of ducks. Individuals need to look in the *zone*, or immediate area, for suitable cover. Once contact has been initiated and the team is taking fire, the team needs to have simple and effective battle drills that can be initiated in the event that verbal communication is no longer possible (gunfire).

Further, these drills need to be in sections or modules so they can be easily remembered. For example, as previously described, a team should have a simple drill that will explain what to do when moving to a door or breach point on a potential target. The drill should generally be the same each time and should be easy to execute. By keeping it simple, your mind can catalog it and prioritize it every time you come into a high-risk situation.

Depending on your career field, breaching an entrance may occur a great deal in your career. How many times do either military or police have to move to the front door of a residence or compound? This simple action must be accomplished on every target of every mission. It applies to hostage rescue, high-risk warrants, search warrants, and barricaded persons. Develop and implement one technique that will work for all missions, and help your soldiers or officers at ground level by requiring them to have fewer and simpler drills, which will allow them to react more efficiently in a high-stress situation as they will have less mental clutter to deal with.

Validating your drill should be a prime concern and focus. How do we do that? Live role players, scripted scenarios, and videos are the key. Develop five worst-case core scenarios that are likely to happen, and develop tactics to solve these problems. These tactics should ensure the maximum safety and survival of your forces while ensuring effective and decisive fire on the enemy. Also, reviewing worst-case scenarios ensures one of two things is likely to happen. First, if you don't encounter the worst-case scenario in training, it may hit you in combat. Second, if you are lucky and don't plan for this worst-case scenario in your first training cycle, you will probably find it later and have to do double the work and adjust the current training or undergo another one to solve this new problem.

A word to the wise: *Make sure that you record the actions and look at it from all angles.* I like to shoot the video from above, so I can see how people move and where their weapons are pointing. Also, by using brightly colored paintballs, you can actually see how the fire is distributed on the battlefield. This can also be done with tracers in a day or night exercise to see how well your men are hitting their targets or how effectively the fires are massed against your enemy. If it is feasible, allow all your personnel to watch the video to gain a better understanding of what the big picture is and their part in it. As is true elsewhere in life, knowledge is power.

## DECIDE

### *Individual Level*

The individual decision-making process should be simple and efficient. The most efficient way to deal with this issue is to simplify and interpret the rules of engagement. The ROE should be designed for the lowest member of the element to understand and react to a threat in an efficient manner. ROE in the military is generally written by staff judge advocate (SJA) personnel (lawyers by another name). Since lawyers are generally too educated to produce a simplified ROE sheet, some ROE sheets can run a page long.

A page of writing in the mind of a private is too much information. The private, sometimes carrying a crew-served weapon such as a light machine gun, can do a lot of damage under the right conditions. Conversely, a private failing to shoot at the right time can enable an enemy to gain the offensive and you to lose the window of opportunity to inflict maximum casualties.

This window of opportunity may be hampered when a private is confused by a complex ROE. The private should have a streamlined ROE that enables him to make a split-second decision that will weigh in his favor. Remember, bad guys do not have a death wish. Most hope to survive the confrontations, as we would. Having said this, they may approach you and not see you until you open fire on them. Either you hit them or you don't, but if you don't, they may scurry for cover, drop their weapon, and jump outside your ROE box. They may decide that hauling ass away from the engagement is their best chance of survival.

Technically a soldier cannot shoot them when they depart, unless they are carrying a weapon. These combatants could be running to get another weapon, but you have to ensure that you have a supportive chain of command in this matter. Should the soldier be uncertain of the ROE and the bad guy spots our soldier first, he may elect to do the same thing and run from the area. The private with his big gun needs to understand when he can shoot and when he cannot to make best use of his weapon system.

A TL may not be present when someone comes into your field of view, and it may be solely up to you whether you shoot or not. That same TL may be five feet from you and does not have your visual angle and may fail to see the threat. Consequently, an individual needs to know when they can pull the trigger and when they cannot.

How do we simplify this process at the individual level? Easy. We take the same basic concept that we use for self-defense in the United States and we transfer it overseas. If you feel your life or the life of another is in imminent danger of death or great bodily harm, you can use deadly force to protect yourself or others. While TLs and sergeants may have to use examples to illustrate this concept to their young men or women, in the end, the decision-making process should be simple. It should take longer to switch a weapon safety to fire and pull a trigger than to make the decision whether to shoot or not to shoot.

These decisions can be complicated today by an enemy's attempt to confuse us by wearing civilian clothes in combat. A rocket-propelled grenade and launcher is not some new fashion statement, rather, it is a deadly threat. The same thought process can be said for noncombatants rushing into a gunfight. They are not coming for some social purpose, but rather to aid the enemy, and they should be considered a threat.

I have witnessed noncombatants bringing ammunition to enemy positions, spot and direct fire for enemy gunners, even haul the enemy's wounded away. In all these cases, under proper ROE, they have lost their civilian status by actively aiding the enemy. Further, if they are not running away from the battle, they are keeping bad company. Once again, these individuals have forsaken their civilian status and should be considered a combatant. If you catch someone looking around a corner without a weapon and they see you, they should run in the opposite direction. If they fail to run and look again, they are spotting for an enemy soldier, and they are plotting against you. The next time they look, they should see your front sight post and your right eye staring at them, then a flash from the end of your barrel.

The goal now should be to condition our soldiers to think the same way and break them out of the "uniformed military" thought process. We have grown up preaching that our enemies will be in uniforms, and when we don't see these uniforms, our forces hesitate in firing. Or once they do fire, they have remorse and second thoughts as to whether they have done something bad. This needs to be addressed before the battle, and the leadership needs to praise and reinforce their belief that they did the right thing.

### *Team Level*

Team decisions should not vary from that of individuals; rather, team actions should be enhanced by an individual's quick thinking. The faster the individual reacts to a situation, the faster the team will maneuver to back him or her up and provide additional fire support. Again, to start the process rolling, it takes the individual to be alert and decisive. Further, decisiveness at the individual level should be rewarded and reinforced and not punished.

### *Organizational Level*

As with the team process of decision making, the organization will be able to react or respond faster if the individuals and teams make faster assessments and decisions. The organization may have other plans at the time and may be waiting for the precise moment to prepare a decisive strike. For example, you may "bait" the enemy in with a feint of sorts and project an illusion of weakness where you then liberally apply large doses of bombs, artillery, and direct fire to his massed forces.

This was an unintended result of actions in Somalia in 1993. When the Somali National Alliance (SNA, a.k.a. Bad Guys) thought that the Task Force Ranger was pinned down, they began gathering their forces from all over the city in waves to come in and assault the defensive perimeter. As they would come down the streets in groups of forty to fifty people, little bird gunships would fly in between buildings and chew them up with miniguns. You would have thought that the Somalis' learning curve would have gone up. It did not, and they did it again and again. They unintentionally threw their people into a meat grinder and allowed the pilots to keep their guns warm. Never before could we get all their forces in one place to engage them in such an efficient manner. The pilots took advantage of this and quickly annihilated the enemy's fighting force.

*Training for Decision Making*

Teaching decision making is a process that has improved over the years through the use of better technology and updated theory. Over twenty-five years ago, law enforcement used wax bullets on a white wall or background while a scenario was projected. We have similar, though better, technology today in the form of visual interactive simulators and live role players.

Interactive simulations can be played with lasers and can provide you with both the decision-making process and the requirement of making accurate shots to stop your opponent. Simulations and live role players can add a bit more stress with a true-to-life picture and scenario of a problem. The scenario can be replayed, or an element added or deleted at a moment's notice.

Simulations should not be limited to the individual level but must also be required of the command personnel. If we are all affected by the same decision-making process during a high-stress situation, why not require a leader (who makes all encompassing life-and-death decisions affecting an entire force) to also rehearse or practice these decisions?

Sadly, there is no requirement for such leadership training. Only extraordinary leaders take the initiative and seek out this training. Generally, most of today's officers feel that they don't need the training, or that it is too easy. Yet these are the same officers who are overwhelmed in the stress of combat while trying to process all the incoming information under chaotic conditions. Today's combat is fast, intense, and lethal—requiring leaders to make positive decisions faster than ever before. The training discussed in this section should be a mandatory requirement for leadership.

## ACT

### *Individual Level*

The *act* phase is where I place most of my emphasis. The most fundamental act is pulling a trigger. As Special Operations soldiers, we spent a majority of our time on the range, perfecting our skills. This was a luxury we took full advantage of. The first three steps in the loop are important in order to reach this stage. Using the act of making a surgical shot at seven yards with a pistol, I will attempt to put this step in the loop in relation to the other steps.

I train to make my first shot from a high ready position, a position where the handgun is close into the body, muzzle slightly up, and generally following my eyes as they scan. As I observe any potential targets, I have already oriented my body to the act; and after I determine that a target is bad, I commit, or *act*, firing the shot. The act itself is pushing the weapon out, taking the slack out of the trigger at the same time, catching the front sight as it comes up on target and dropping it into my rear sight, breaking the shot as this happens. I practice so that I can do all this in one second or less and hit a six-by-twelve-inch kill zone resembling a spinal column on a target at seven yards. I have refined my *act* phase to accomplish the task in less than one second, routinely attaining the .90-second mark.

How do you get there? Repetition and muscle memory.

Top shooters practice a dry run of shooting 70 percent to 30 percent live fire. In other words, for every time you fire live rounds, you need to practice dry two or three times. You must also understand that with muscle memory, at least two to three thousand repetitions are necessary to help master a simple skill. So, at night, on your own, you must practice with a timer to ensure you are making the time standards. Occasionally, you should use a video camera and record your sessions to ensure that you are not practicing or acquiring any bad habits. This is a constant cycle. Dry practice is the key.

There were many times in my career in Special Operations that while in threat areas, we did not have the luxury to practice and live-fire our weapons. Instead, we practiced fifteen to twenty minutes a day against a safe wall with an empty weapon, reaffirming all the motor skills required to accomplish the act.

### *Team Level*

The team needs to implement the same protocols as the individual, but at the team level. The TL should ensure that all members are familiar with the team battle drills to the point that these drills are ingrained in them. Begin with a class on the drill, and then move to an area where you can dry practice. Once you have the concept down, move to simulations or paintball and live role players. Then take the drill to live fire. The same training problems apply in a combat environment, and that involves the time and area to practice. Team leaders can have their teams do dry practice by walking through buildings or garages on a reduced scale in their down time.

### *Organizational Level*

As suggested before, organization leaders should constantly work at perfecting their leadership and decision-making process. They should observe and become involved in the lower-level training to ensure they are aware of what their soldiers are doing and what tactics they are using. Further, they should seek to understand what actions and decisions they make to help the teams accomplish their mission. Also, they must fully appreciate the team's capabilities and limits. Finally, they should understand that it makes more sense and is more efficient to let the team complete the mission rather than to micromanage. Only when a TL demonstrates their incompetence should an organizational leader step in and take control.

## CONCLUSION

Boyd felt *orient* was the most important phase of his cycle. It is important to have a tactic ready that is oriented to the particular situation you are in. But I feel that if you fail to see or are hesitant to decide on a course of action, you can encounter problems. While continuing to perfect the *act* phase, you must not neglect the other equally important phases of *observe*, *orient,* and *decide*, which will cut your time navigating through the loop. Acting is important, but as much time can be cut from each of the other phases as the *act* phase. Equally important is the ability to rapidly "see" the enemy in the *observe* phase and to have aggressive tactics or techniques ready in the *orient* phase. Finally, the decision-making phase should be efficient to ensure that the act can be carried out swiftly and decisively. I prefer to cut time in all phases of the loop while training as an individual or as a group.

## KEY POINTS

- Understand Boyd's Loop and work to save time and increase efficiency.
- You must see before you can shoot.

# PART I

## THE MENTALITY AND ACTUALIZATION OF LEADERSHIP

# 1

# COMBAT MIND-SET

*Take it like you own it and leave it like you sold it.*

—From a former Special Ops sergeant major

- **COMBAT MIND-SET DEFINED**
- **THE PROBLEMS WITH HUMAN NATURE**
- **DEVELOPING A SINGLE COMBAT MIND-SET**
- **AGGRESSIVENESS IS THE KEY**
- **TACTICAL CONFIDENCE**
- **INDIVIDUALIZED MENTALITY: FIGHT- THROUGH MIND-SET**
- **TEAM LEADER MENTAL PROGRAMMING**
- **TACTICAL COMMANDER MENTAL PROGRAMMING**

## SCENARIO: NIGHT ACTION

We had a full moon, one that seemed to be one hundred percent illumination. I watched a couple of Rangers move from the command post (CP) to the corner of the RPG alley with what appeared to be an M60 machine gun. They hadn't been there more than a few minutes when a gunman fired an RPG that impacted the corner they were using for cover. The impact of the round and the subsequent explosion knocked them off on their ass. A few folks from the CP came out and dragged the men back to safety. I called in to the assault commander (AC) and told him to get the gunships in and work the alley over. He reported to me that the command and control (C&C) bird was not sure where everybody was and that they would not give us fire support.

I was fucking pissed.

I told him that there were not any friendlies up that alley. For a moment, I lost my composure and switched the frequency on my radio from the working net to the command net that the C&C bird was monitoring. I called my commander in the air and told him, "You get those goddamn gunships in here right now." I relayed that I just watched two Rangers get peeled off a wall and that, "there were not any good guys up that alley."

Afterward, I switched my radio directly to the gun bird pilots and requested fire missions. They were more than happy to deliver some steel on target. Actually, they had already taken the initiative and performed some gun runs without the C&C bird's approval, keeping many of the bad guys off of us. These pilots set the standard time and time again, proving that they had the greatest courage, initiative, and discrimination, culminating in an incredible warrior ethic. The gun bird pilots requested that we mark our position, and we did. Shortly afterward all the positions were marked. Anything outside our markings was fair game for their miniguns and rockets. They continued to tune up anyone intent on causing us problems.

Things were starting to slow down as the sun set. We got quiet and started using the shadows of our new home. You could hear the pilots servicing some large groups of militia who were assembling and trying to move on us. The bad guys were bringing their forces from all over the city to rendezvous points, issuing battle plans and then moving toward our positions. The pilots were engaging groups of forty to fifty people at a time, clearing the street with their miniguns, which reminded me of a chainsaw biting into wood.

Soon the smaller enemy began to probe us using three- to six-person elements. I was scanning the intersection twenty-five meters to my left and then back to the corner of the alley about twenty-five meters to my right front. It was quiet – too quiet.

Suddenly I heard loud jabbering and saw three individuals appear out of the alley to my right. They were dressed in dark pants and light-colored shirts. I asked Tony, my assistant team leader, if there were any Rangers still in that position. He said, "No."

I raised my CAR-15 from its low ready position and swept my safety to full auto. I placed my tritium front sight post on the middle thug and knew that I would sweep from center man to right and then back across the cluster. I figured Jake would take the lead guy. I braced the heavy CAR-15/shotgun combo against the doorframe for added support. I waited a few more seconds for Jake to take the lead. Jake was twenty yards across the street in the CP, ten yards from the alley corner. Jake was on them and illuminated the first bad guy with his white light.

The man stood there stunned for about a second or two with an AK in a low ready position. They looked like raccoons caught red-handed in a garbage can. Jake began to service him with a few rounds of 5.56 green tip (standard issue NATO cartridge). I cut loose into the center man full-auto and swept to the third and then back to the first, firing a total of about fifteen rounds.

Out of my peripheral view, I could see the rounds sparking off the alley wall twenty feet behind them in a tight pattern. The weight of the weapon and the supported position helped manage the recoil and keep my group tight. The next day showed a group slightly larger than a basketball in the wall behind where they had been standing.

The gunman farthest to the right in the group went down hard and fast, while, to my amazement, the first and second man moved around him and began dragging him back up the alley from where they came. The boys across the street said they did not make it far up the alley. They later reported that they could hear moaning probably where the trio collapsed and died from loss of blood.

Tony said calmly, "We should probably spread out." I began to respond, "That's probably a good idea." We had taken two to three rooms of what might be considered a small house or apartment. Tony took Kim and moved through a barred window that Scott had torn out. They linked up with a fragmented group of Rangers and a couple of our guys to include a Special Operations medic. This window separated the house we were in and the next house, which connected to the alley and overlooked a crashed Black Hawk helicopter. As a buddy team, they each took a window with different fields of fire, but could see each other in the same room. One controlled the street to the north and the alley where the bird lay, and the other controlled the alley to the west where we had taken so much fire earlier.

Across the street, the CP had gotten quiet. We had several leaders there, four as a matter of fact. The AC and another leader of the same rank, plus two senior NCOs and an assault team were present in the courtyard and in the house. The second officer was along on the hit for a bit of on-the-job training, while the two senior NCOs ran platoon-sized elements. From my perspective, they were all relying on the one AC to absorb all the incoming information, process it, and then make all the tactical decisions. I think these "leaders" were in a bit of shock at how quickly and violently the battle had escalated. Probably 80 percent to 90 percent of the information was going from them to the C&C bird and not to us on the ground.

At the TL level, we were trying to tie in the now current positions of our perimeter with interlocking fire. I was on the internal channel talking with the other teams, trying to establish fields of fire and trying to ensure that we had all the approaches covered. It is tough to do at night, talking about intersections and terrain features, hoping that you and your counterpart were looking at the same area. We established a protocol of sorts: When you were about to shoot at incoming enemy, time permitting, you would radio and alert the force. You gave a direction, description, and distance; and then you would engage your threats. You would then give a brief call when you finished shooting to let everyone know what the outcome was. Sometimes the bad guys came in too quick, and you just had to shoot first and then do your call. The first method helped ease your nerves, because you knew in your mind what was coming. The second method caused you to tense up until you received the status report. You did not know if the team or your position might be overrun and you might have to turn your direction elsewhere. Also, by being alerted, you could tuck back in and not be exposed to friendly or enemy fire or "bleed over" fire from the exchange.

During this and other actions, our internal radios enabled us to effectively communicate and keep the force informed. The one technical problem we faced was that of battery life. As most law enforcement personnel are aware, communications are one of the biggest weaknesses in the system. Ours were no different. Our batteries were a rechargeable type that did not last long when transmitting. For years we had asked for a disposable lithium-type battery that you could carry in an emergency and talk on for days when things got tough. To ensure constant radio communication, one person on the buddy team would turn their radio off and receive the information verbally from his partner to save battery life. After a few hours, they would switch radios, and the other man would turn his on. This went on all night.

Everyone was on edge, and my biggest fear was that of being overrun. I went through a mental checklist of my equipment, my "layered offense" as I termed it, and how I was going to do business if the bad guys came at us en masse.

I checked all my rifle magazines and ensured that I could easily get to them should we get hit with a wave of enemy bodies. I thought about stacking them on the window ledge next to the door, but that would limit my mobility. I would be stuck there in that position, good or bad. I chose to keep them on my body. I checked my shotgun rounds; I still had the thirty or so I had brought in with me. Should my rifle fail or I run out of ammo, I would go for my shotgun and #4 buck. Should I need to transition from my shotgun, I would go for my pistol.

I had one mag in the gun and two on my belt. My final check was my last frag and my knife. I had one large fragmentation grenade left. I would save my frag for a large group of fighters or to clean some hard cases from out around a corner. My knife was for when it got up close and personal. Before I got to that point, I would prefer to pick up an enemy AK assault rifle, as I always had a soft spot for it in my heart for its reliability and knock-down power and for the power of its cartridge.

As the night wore on, Tony and Kim started to get some trigger time. They would engage the bad guys here and there, allowing them to come down a wall to an indefensible position and then open up with a 40mm grenade to the front and 5.56 on both sides with great results. At one point, I was scanning the intersection to my left, one that was supposedly covered by a sister team, when this bad guy comes walking down the center of the street. I was trying to put my light cover back on my rifle gun light when I had a white light accidental discharge (AD).

Simply put, I screwed up, and my white light flashed the ground. Instead of going straight, this recon scout made the last bad decision in his life and turned toward us and started walking down the center of our street. He was doing a recon, trying to pinpoint our positions so he could later bring back his friends with RPGs to try and root us out. I pulled my pistol out and started tracking him, putting my tritium sights center mass of his right side. He was a big guy, over six feet and stocky. Looking at the background of my target, I realized would be shooting almost directly into the CP and my muzzle flash would be exposed to three different directions.

So I got on my radio and called Jake and told him to take the guy out when he leaves the perimeter. Jake had a great position at his gate that concealed him from all angles but one. Jake acknowledged and let the guy walk about fifteen feet and then fired one round that struck him in the lower left of his back and exited the right front side, the bullet poofing out of his shirt as it exited his body. The guy spun around and looked at Jake for a second, at which time Jake serviced him with two to three more rounds in the chest, dropping him in his tracks. This bad guy dropped in the wrong spot and later became a speed bump for some of our recovery and convoy vehicles.

Immediately a voice came over the radio, one of the senior NCOs in the CP, saying, "I don't think he had a gun." I thought to myself, "What a dumb motherfucker." He still had not switched over to combat.

Just then, as I was scanning back to my left, I saw another guy was walking down the center of the street where the first one had come from. I raised my rifle and was tracking him with my front sight, and I had a good squeeze going on my trigger, when bam, the guy dropped. A sister TL (who I thought had that area covered) had just opened a window and saw this guy in the middle of the street, probably twenty feet away. He tagged him with several rounds and immediately got on the radio, saying, "You gotta tell me when these guys are coming in." I laughed and told him that I thought he had that area covered. He did, and I did not take it for granted. Things tend to look different in the day than at night. Lesson learned.

An hour or so later, the two deceased recon scouts' friends decided to come pay us a visit. The light from the moon was so bright it produced night shadows. These were cast from buildings, trees, etc. I was watching the intersection to the left again, even though it was supposedly covered, and caught sight of a Somali gunman on a knee, poking his AK around the corner. He fired one shot down the wall toward the area of the CP. I told the guys in the CP to tuck in and hold tight. The path leading to the CP was sloping and the round went way above their heads. This guy was reconning by fire. This is an old military technique where you shoot and then see who shoots back.

We all held our fire, and he got a little braver. He crawled on his hands and knees halfway down the wall with his AK tucked under his arm, moving like a jungle cat, slow and precise. He used the shadows as he moved, and I viewed his movement as a work of art. He moved up to a point about twenty feet from the CP, looked hard for a moment, turned around, and went back to the corner from which he came. I told Scott to get ready—we were going to get some business. We had just taken up a high/low position when all of a sudden the corner where the bad guy was exploded in gunfire, one-sided, of course.

It seemed the other team spotted him, and there turned out to be five more of his friends getting ready to move on the CP. Our sister team worked them over good, putting rounds into all of them. Again, they managed to drag their dead and wounded back up the street. Once the first light came, there were no bodies to be seen.

## AFTER-ACTION COMMENTS

### *SUSTAIN*

- Continue to require individual initiative at the team member and TL level to ensure that they are thinkers that are shooters and shooters that are thinkers.

- Ensure the communication process continues at the individual and team levels.
- Know when to change from a surgical mind-set to a combat mind-set. Discuss this before you go into harm's way. Otherwise, as a leader, let them know when to shoot by setting an example.

*IMPROVE*

- Don't layer fire support. The soldiers on the ground know who is shooting at them and from where. Generally, fire support gets screwed up when you get a third and fourth party involved.
- Senior leadership needs to get proactive; check the perimeter and support the teams in the fight. Again, lead by example.

## COMBAT MIND-SET DEFINED

An aggressive combat mind-set is possessed by people who can screen out distractions while under great stress to focus on the mission and are willing to go into harm's way, against great odds if necessary. Hemingway might have described it as "grace under pressure." Simply put, we must be able to maintain our focus and composure and not allow fear or stress to cause us to make stupid mistakes. Combat mind-set sets the stage for all components of this book. Without it, you will be unable to employ positive and decisive leadership in critical situations.

The problems that affect combat mind-set lies with human nature. When the sound of shots ring out, the average person will stop and cringe physically and mentally. As soon as their mind registers what the shots actually are—i.e., gunfire—their fight-or-flight response will kick in. For the average human, the urge is to flee.

Special Operations soldiers, including law enforcement tactical personnel, must act contrary to human nature and must control their fear and channel it into *controlled aggression* and move around, and sometimes into, the fire. This even applies to the average police patrolman on the street responding to the report of an active shooter call, where a gunman is shooting innocent people as though they are sheep. Military, law enforcement tactical personnel, and patrol officers need to be mentally, physically, and tactically ready to wade into a fight, bypassing injured and dying innocent people to quickly and efficiently neutralize a threat so no more innocent people will die.

The ability to develop controlled aggression is dependent on a person's ability to channel their fear, anger, and anxiety into a focused mental package. Channeling and controlling this energy is routinely what military individuals term as *high speed*. They use it to describe the caliber, efficiency, or speed with which a soldier or team operates.

High speed in my dictionary is the ability to apply the basics on demand the first time in a high-stress situation. Some people falsely think it counts and can be accomplished after a dozen rehearsals. The real people who wear the high speed badge can execute a live-fire explosive breach cold hit on a target with an unknown floor plan and not kill or injure themselves or any innocents.

More times than I wish to count, I have witnessed individuals with uncontrolled aggression screwup missions, tactical problems, and routine combat actions. Instead of donning their equipment for the right reasons, with the understanding that it takes thousands of hours to attain a basic skill level, some Special Operations soldiers wear the gear for the CDI (chicks dig it) factor and forget the reality that this is hard work.

Success will require countless rehearsals and constant maintenance training to hone individual and collective skills. Some individuals tend to rely on their physical prowess to solve tactical problems, leaving their brain somewhere far behind. Combat is a thinking man's game, and you need to rely on skill versus luck to ensure your survival. *No matter what your problem is, you have to get your mind right first to ensure your survival.* If you find luck, embrace it, but don't rely on it.

## THE PROBLEMS WITH HUMAN NATURE

The mind-set I choose to use throughout my career has served me well. Through the years I was able to watch fellow soldiers and law enforcement individuals and see what worked and did not work. I grew up in an era of Vietnam veterans who recently separated from service and entered the law enforcement arena. They were a great crew—confident, professional, no-nonsense guys who did not take crap from lowlifes or dirtballs.

The other side of this coin are the shit talkers I have run into I have come to the conclusion that braggers and loudmouths are the first to crumble in a high-stress situation. I venture to say that their outward projection is a smoke screen for a lack of confidence. I will not generally lump them into the "coward" category, but they are close. I have seen too many shit-talking leaders crumble when the first bullet snaps by their head. Their balls shrink from their self-perceived bowling ball size to that of BBs. I equate loud talk with foreshadowing of failure. This is my mathematical view of boasting:

> Loud talk = failure = impotence/cowardice

As for egos, leave your ego behind in your wall locker; or better yet, permanently deflate it. Learn to focus on the task at hand and solve one problem at a time. Should you make a mistake, admit it and concentrate on how to fix it. All an over-inflated ego will do is confirm you are an asshole and make you look like a pussy when you crumble under the stress of combat. Don't talk about doing something—do it. It is the only way to make believers of nonbelievers. An old martial arts instructor once said, "Believe half of what you see and none of what you hear." There is much truth in this. Unfortunately, I have run into too many individuals who talk a good game but cannot perform to their level of boasting.

As a final point in this section I would like to tie in a final point General Patton mentioned in his book *War as I Knew:*

> *If we take the generally accepted definition of bravery as a quality which knows not fear, I have never seen a brave man. All men are frightened. The more intelligent they are, the more they are frightened. The courageous man is the man who forces himself, in spite of his fear, to carry on. Discipline, pride, self-respect, self-confidence, and the love of glory are attributes which will make a man courageous even when he is afraid.*

The point I would like to make is that the more one develops one's mind and situational awareness, the more control one will need in combat to sort out all the information being processed. Call this courage or call it a *combat focus*; it is something you will need to help you process the combat information that your senses absorb.

My idea of a combat focus is twofold, consisting of a soft and a hard focus. Our soft focus is a relaxed vision that I routinely operate in and consists of looking at the big picture, a landscape, so to speak. The hard or detailed focus is where you pick out a point on that landscape, such as a bush or a rock to focus your attention on and "see" a particular point. Combat and close-quarter battle (CQB) requires the ability to change from soft to hard focus and back again almost instantly.

In combat, you are continually scanning the environment or field of view for threats. When traveling down a street under fire, you're looking for the obvious—people with guns, muzzle flashes, etc. As you scan for the obvious, you start looking hard at places that can conceal a shooter—dark windows, doorways, and rooflines. You are continually going back and forth through this visual process. You might see a flash from a window, zoom in on it for a second, put a few rounds on it, and then pull back to a soft focus of the entire terrain again. You might go back to the window again, just to ensure the threat does not reappear.

Let's add something more to this picture: When moving and shooting, imagine bullets are also coming your way. Some bullets might be going high and you just hear the report of the weapon, other rounds may zip by your head and snap or crack as they go by. Others still may spark the ground around your feet. This is where Patton's comments hit home for me. I know that as I became more aware of my surroundings my mind attempted to absorb and process all this information.

The key is to sort out what is critical information and what is not. I found that the "air balls" were nothing to get excited about. Yes, someone is shooting your way, this does happen in combat. More likely they are either a bad shot or just slinging lead. You need to go into combat expecting people to shoot at you and not be naïve about this danger.

The rounds that should be getting you excited are the ones that are cracking by your head or sparking the ground next to your feet. These should be sending a signal to your brain saying that someone is zeroed in on you and doing their best to kill you. Your options are simple: Seek a better route to get to your destination, or get some cover. Sending enough lead your way, the bad guy will eventually hit you if you continue your same approach. Even if they only hit you with one out of ten rounds, it will still hurt.

## DEVELOPING A SINGLE COMBAT MIND-SET

I know I have said this before, but I will hammer this point home again: *You must firmly believe in what you are doing and why you are doing it.* It can be for your country, the organization, your team, or your buddy next to you. It can be that inner drive that says don't quit and do your best. *Whatever motivates you, you need to harness it and keep it strong in its place.* Reflect on it as needed to keep your energies channeled for the time that will come for you to earn your keep. The stronger your belief, the stronger your mind-set.

This resolve or strength will also help ensure your survival. With it, you will train harder and push farther than someone who does not have it. Use this strength to develop your own personal beast and then keep it in its place.

Some folks may look at me funny when I use this analogy, but it is simple. When you are confronted with a desperate and possibly lethal situation, your physiological and mental system kicks in. Breathing, heart rate, and adrenaline all come into play. Your fight-or-flight response also factors into the equation. The key is not to figure out the best way to run away, but to figure out the most efficient and violent way to remedy the problem at hand. Learning to channel the energy into positive action and thought is the real key.

Develop your personal "beast" for these times. Everyone has that switch they can throw that will take them to another level or fighting plane. Go to a place where you can become emotionless and totally focused. You must have the ability to be aware so you can deal efficiently with multiple threats and be successful. Still keep the emotion harnessed for that little "umph" you may need should the situation become desperate.

I was able to develop this extra drive at a young age, and then I harnessed it for future use. You can use it while doing live-fire CQB, ramping yourself up to a mental level where you are conscious of everything going on around you, explosive charges going off, shooting, shotgun breaching, flashbangs, assaulters screaming and putting people down in the next room. You can learn to focus through this and do your job and turn the emotion on when you need to. Your job may be as simple as clearing a single empty room or as complex as dealing with a room containing multiple friendly innocents who are bleeding and screaming. You must maintain your focus and composure while swiftly and efficiently putting surgical rounds into the right target.

One mind-set that I firmly keep was of the *layered offense*. Everyone and their brother wants to talk about being defensive. Generally, defense does not win personal or collective battles. In my mind, I always wanted to stay focused and in the fight. I used the knife on my belt as a cutting tool, but it served a dual purpose. I would mentally program it as part of my layered offense. I had my rifle to employ as my primary weapon. If it malfunctioned or I was out of ammo, I would go for my pistol. If my pistol was out of action, I would try to pick up an enemy's weapon. If none was available, I would then go for my knife.

The knife I carried was virtually indestructible, and I kept it sharp. The only part of it that could fail was the person wielding it—me. I knew deep down that I had to maintain my belief, my skills, and my physical condition to survive. This kept things simple and in perspective for me. Fortunately, no one had ever made it past my rifle in combat, but should I have needed to, I would not have hesitated to transition to my next weapon system. This mind-set could be termed a weapon loop, where I always knew where to go and find my next lethal system to employ. Keeping it simple kept me fluid and efficient.

Once I developed this layered offensive mind-set, I had to weigh in being fast and sloppy or methodical and precise in my tactical ways. I chose the latter, erring on the side of being methodical and precise versus fast and sloppy. Instructors I had in the past always reminded us that "smooth is fast." We were continually pushed to be precise and in control.

Furthermore, I developed one mind-set for all tactical situations. Special operations and law enforcement tactical teams usually have four or five standard high-risk missions they are tasked to perform. In the law enforcement arena, these are hostage rescue, high-risk warrant, search warrant, and barricaded persons. What is the constant for each mission? They are dangerous, and you can get shot at while doing them.

I recommend that you develop a simple and aggressive mind-set that will work for each mission. I developed my personal combat mind-set to help me deliver efficient lethal force on demand. I went into every mission with the belief that someone was going to shoot at me, and I expected it. Failure to do this will leave you in a mentally unprepared condition to deal with the violence of action that someone might bring upon you. You will also be too late to ratchet up your aggressiveness once the bullets start flying.

*I am going to fucking destroy you*, is the thought I firmly and quietly kept in mind, along with a game face of focused determination when I was preparing for a combat mission or doing rehearsals with role players. I treated both rehearsals and combat the same way for simplicity's sake. As for role players, they come in two types, professional and standard.

The professional role players are the ones that have been role players before and like to try to get the drop on an assaulter or good guy. Some will push you and try to control a situation during an assault, creating additional chaos for you to deal with. This can be as simple as acting out a script of a panicking passenger or a smart-ass that wants to verbally challenge you during the scenario.

It is important to gain physical, emotional, and mental control of the situation as rapidly as possible. Once you get security and neutralize any obvious threats, you need to immediately deal with these kinds of people. Failure to do this will make your job much more difficult. If you "educate" or "tuneup" the problem child as rapidly as possible, especially in front of his peers, you will make believers of nonbelievers.

## AGGRESSIVENESS IS THE KEY

I usually begin my combat mind-set module with a talk about aggressiveness. This is the key to success.

ACT versus REACT

Your mind-set and attitude should not be

- Passive
- Reactive
- Neutral

Your mind-set and attitude should be

- Pro-active
- Aggressive

I teach in my classes that there exists a pool of aggressiveness in any combat situation. Either you will take it and use it, or your opponent will. The choice is yours. Chaos exists in the battlefield and has more effect on the mind and perception than on reality. Chaos can be used to describe the combined sensory overload that affects you during raids or combat operations. Our job is to bring order to chaos. Sometimes we need to bring in our own brand of chaos to help establish order.

Imagine, if you will, an environment with brilliant flashes and the concussion of explosions going off near you, people shooting, doors being shotgun-breached, distraction devices flashing and booming. This is chaos—induced chaos. This chaos helps you to do your job safely and efficiently. You must learn to understand it, be aware of it, and let it pass by your senses. You use it to control everyone on your target. Once you gain control, you can ease back as your target audience dictates.

In a hostage rescue scenario, everyone will be physically, emotionally, and mentally dominated from the instant they lay eyes on the team. I want them scared to death upon our entry, with individuals curled into a fetal position. This quickly sorts out who is serious and who is not. Fetal positions are a good indicator of compliant hostages. I joke wih my students that they can always get therapy later, but they have to be alive to get the therapy.

Most of the time, the chaos you bring in will take the fight out of the most hardened opponent. I remember other men recounting instances where all the bad guys on the target were stretched out prone after the first explosion went off. They did not want to play against guys who were serious about doing business. I also remember a raid where the father had his family lined up in front of the breach point and held their infant out as a shield in front of them. This happened not once but twice during one operation. These guys were supposed to be high-ranking officers in their militia. In effect, they were low-life cowards. Americans wonder what separates us from other countries. It's how we cherish and protect our children.

Occasionally, you will launch an assault on a hard target, where no one will throw their hands up and surrender. For this, I always believed in developing and going in with one mind-set—the mind-set that it is going to be a hard fight. I always prepare mentally for that one guy that is going to fight back and try to fight through me, the guy that I am going to have to shoot to pieces in order to stop. It has happened before, and it will happen again in the future. Law enforcement officers occasionally run into individuals with a fixed combat mind-set, and they literally have to shoot these folks apart to get them to stop their aggressive actions.

## TACTICAL CONFIDENCE

By developing simple and effective drills that will apply to a majority of your missions, you will develop a confidence that is instilled from the lowest team member to the TL. Why is it important? Simple: If your men do not believe in the tactics they are going to employ, they will generally fall apart once the shooting starts. Also, if your tactics are based on the best-case scenario, versus the worst-case scenario, your men will have too much to deal with mentally once things start going wrong.

All of us have the talent, expertise, experience, and resources to develop simple and effective battle plans that will work on all missions. For example, moving from a drop-off point from a van or moving to a target on foot is basically the same wherever you go. If someone is going to shoot at you, your tactics should address this, and it should work for all missions. Simply put, getting shot at is getting shot at. It is the same whether you are on a hostage rescue mission, a high-risk warrant, a search warrant, or barricaded person–type mission. For all these missions, I teach a simple drill called Tactical Flare, where the men move into positions that spread them out, away from the gunfire, allowing them all to shoot safely and bring all their firepower to bear on the threat.

Generally, I see movement formations to the initial breach point of a target consisting of a line of ducks where only the first man can safely shoot. Should they encounter a suspect or take fire, the team relies on the lead officer to solve the problem. This is in effect a one-on-one gunfight.

For starters, I put a minimum of two men up front with two guns and two sets of eyes for a shorter response time. Once verbal or nonverbal communication (shooting) is initiated by the point personnel, the remainder of the team deploys into offensive firing positions.

This does several things. First, it encourages men to respond aggressively forward and to not react, or to maintain a neutral "wait and see" posture. Next, it gets the team spread out and out of the bad guy's cone of fire. Further, it mentally overwhelms the threat because now he is not facing one lone officer; he has multiple aggressive movers to contend with as well as several weapons pointed at him. You create a *reactionary gap* in the subject's mind, and then you exploit it. There is nothing like good old-fashioned aggressive action.

Using this technique, the officers will have a better visual angle, and should the officers decide to shoot, they can all do so safely. I learned long ago that more guns will make the bad person go away faster, either mentally or physically.

Let's throw a variable into the above scenario: An officer has just been reported down. If an officer gets shot during the drill, what changes? Nothing. The team will neutralize the threat and then recover the officer once it is safe to do so. I see too many times where officers try to strong-point the downed officer. By doing this, they become a bullet magnet, and all the bad guy has to do is shoot into a clump of officers. By retreating, they will also have to deal with the threat again. I hate paying for the same ground twice. In my view, it is safer and more efficient to solve the problem the first time and recover the officer once you own the turf.

By fighting through the problem, we have one drill that will work on all missions. It is simple, safe, easy to learn, and does not change from hit to hit. The key point is who we should be designing this drill for. Should it be the twelve-year Special Ops, SWAT veteran, or the new member of the team? It should be the newest and weakest link on the team. Keep it simple. We now have a drill for exterior movement with few variables. Once this is practiced and mastered, we then need to focus on developing a simple drill for your breach point, hallways, rooms, etc. By keeping it simple and easy to remember, it will transition to other tactical situations with ease.

Developing scenario-based drills will also help combat the fear of the unknown. Fear of the unknown is one of the biggest problems encountered when training new team members. Their minds race with all the possible scenarios that they can run into. It is controlled by training on realistic worst-case scenario contingencies and having a plan to deal with them. First, expect to be shot at and have the confidence that your basic battle drills will adequately help you handle the threats. Next, make fear of the unknown a nonissue by knowing what you're going to do in a positive, aggressive manner. This is termed anticipation mind-set. For example, once officers practice the flare, I run five or six different basic worst-case scenarios with role players and see if the drill will effectively manage them. I record each drill and review it in a classroom environment. We then practice our problem areas again and smooth out our actions.

I learned long ago when I was deployed overseas that there will be many times when we are going to come across people with guns. We can't shoot them all, but we can identify the ones that plan on using them against us. These are the real threats we need to be concerned with, and we need to have a plan for dealing with those individuals. I know that action is faster than reaction, so start by developing a *tactical package* that will give your team a tactical confidence and edge in how they perform a mission and deal with these individuals, both mentally and physically.

## INDIVIDUALIZED MENTALITY: FIGHT-THROUGH MIND-SET

There are still a few more points that need to be added to your survival toolbox. First, believe in the cause, in yourself, and in your team. Probably one of your greatest fears is letting down your buddy or your team in combat. A proper mission focus, along with having the proper tactical confidence in your battle drills will help eliminate this fear. If your tactics are sound and you believe that they will keep you relatively safe while delivering devastating fire to your opponent, you will have less fear and apprehension when going into harm's way.

Next, there is nothing wrong with being scared. It is natural. How can you channel that fear? Easy. Look at the technical aspects of the battle and do the math. For example, let's look at getting shot.

If you have done any amount of range fire, you know how hard it is to hit a stationary target, let alone a moving target.

You must first ensure your rifle is zeroed and then practice. The same rule applies to the bad guys, and they generally don't have the training and marksmanship abilities you do. Next, if you use cover, it makes the bad guy work hard at getting an accurate shot at you. So use cover and decrease your chances of being hit. Most importantly, develop a fight-through mentality.

Do not dwell on dying. Focus on your training and what you are going to do to ensure your survival. So you get hit. Big deal. Look at the statistics of those dying from gunshot wounds. They are not that impressive. Generally, you can reach some type of definitive care in a few minutes, or it can reach you. Before you reach this care, who is the best person to start treatment on you? You are.

Start by getting your mind right and being pissed off that some low-life turd shot you and focus your mind toward survival. Do this by ensuring the bad guy is dead and then start your self-aid. As a realist, I carry medical scissors and several bandages readily accessible on my belt. I know by experience that I can get shot by the enemy or by friendly personnel. Should this happen, I will do my best to fight through, neutralize the threat, and let my buddy know that I am hit. I am then going to expose my wound by cutting the portion of my clothes around it and come to grips with my own injury. I will pack it or tie it off with a tourniquet if time permits so as not to take team members away from the team and their ultimate goal of securing the target. I know they will come back for me once the area is secure. The more that I can do to treat myself, the less they have to do for me. Implementing this thought process will start kicking your survival mind-set into high gear.

You can always take the other approach and roll your eyes back in your head and go toward the light. I prefer not to take this approach. You might make it sooner than you expect. When training with simunitions (a type of non-lethal ammunition used in training exercises) or paintball and you get hit, fight through. Why? We have trained our soldiers for years to fall down, quit, and give up once they are hit during training scenarios. This is self-destructive and will cause you problems in combat. You're actually training soldiers to give up at the slightest pain or discomfort.

I have witnessed soldiers with minor wounds mentally shut down in combat because they have trained that way during years of battle drills and rehearsals. For medical training, you can induce casualties, but do it by telling a soldier ahead of time that he is part of a medical scenario. Never allow a soldier or a law enforcement officer to quit on their own. This will start a bad habit that may cultivate hesitation or result in a soldier or officer giving up in the heat of battle, where they generally could have survived. *Finally, if you think you're going to die, get pissed and plan on taking some of the bad guys with you. Do your best to ensure this happens.*

## TEAM LEADER MENTAL PROGRAMMING

The TL should strive to cultivate and demonstrate the same mental attitude as an individual and project this to the team. Your job as a TL is to *live the example* for all to see and aspire to be. We are not talking about perfection, but darn close.

The individuals you are going to lead into harm's way need to have the confidence and belief that your priorities are right. Your priority is to fight smart and hard and bring all team members back at all costs. The team should know this, and it should be talked out before engaging in any mission.

The TL is the first line of combat leadership. If he has a target of opportunity, he engages it in a rapid and efficient manner, as would any member of the team. This may not sound like much, but it can make a world of difference in combat. If your team has not been bloodied together, or been bloodied at all, they may be hesitant to shoot when the time comes due to complicated ROE.

Your status as a new leader could cause this hesitation among a team. A single act of aggressively engaging the enemy will pass as a wave of silent support to your team. You mean business, and now it's time for them to get some trigger time. If I was a new team member and I saw my TL engaging targets, I would feel a slight bit of guilt that I was not doing my job. When you see the boss shooting, you know things are serious, and you should be doing your part.

S. L. A. Marshall wrote about this type of action in his book *Men Against Fire*. It is an incredible piece of work. One incident rang true when Marshall described a leader that would come along and yell at his men to shoot, and then he would leave. Once this leader had left, the men would stop shooting. He described another leader that would come into position next to his men and start shooting at enemy targets and talk to the guys. The men would start engaging targets, and they would continue to do so even after their leader had left. I can attribute this action to two reasons. First, if my boss is firing, it must be serious, and I better do my part. The other reason might deal with today's civilianized combatants and the question of whether it is right to engage people in civilian clothes. Again, if the boss is doing it, they must be bad guys.

On the other side of the coin, each Special Ops person is required to be a shooter and a thinker. I can remember not shooting more times in combat or high-risk confrontations because it was not the right thing to do at the time. We were raised to be surgical and actually avoided more fights than we got into, because it was the smart thing to do.

I remember one incident in a foreign country, when our vehicles were surrounded by fifteen or twenty pissed-off government troops some with fingers on the triggers and the safety off, muzzles stuck in the windows of our vehicles. It was not in our best interest to escalate, so we remained calm during the encounter and waited for the situation to de-escalate. Had one of the government troops had an accidental discharge because of their poor weapon handling, it would have been a different story. We would probably have won the fight, but we would have had casualties because of the proximity of the opponents and number of guns involved. This is where you have to leave your ego and pride in the rear. They may be booger eaters, but they are booger eaters with lots of guns.

In addition, the issue of combat mind-set may come into question when a seasoned team inherits an unproven TL. All the TL can do is live the example and set the standard in all trainings and combat. You should do this as the opportunity presents itself, and you will earn the respect you need to effectively lead your element. You need to earn the team's respect every day and continue to pay your dues. The team, if mature and professional, should not hold it against you that you were in the wrong place at the wrong time and did not see any action in the past. Generally, the team will watch you and your decision-making process. If your team determines that you cannot make a hard decision in peacetime or in training under no stress, they will conclude that you are spineless and will crumble in combat or in a high-stress situation. They are probably right.

## TACTICAL COMMANDER MENTAL PROGRAMMING

The mind-set or attitude of the tactical commander (TC) should generally stay the same as that of the individual and the TL. Hopefully, they have the same aggressiveness as the lower-ranking leaders, but things have a tendency of changing. Routinely, Special Ops leaders at this level, the troop level, are given only a short time to work with and interact with the soldiers. Generally, two years, and they are off to another assignment, usually administrative and as far away from the shooters as possible. Two years is generally enough time to learn the missions and standard operating procedures (SOPs) to become an effective and integrated part of the machine.

Much of the problem with commanders' combat mental programming is that they are rarely faced with the immediate threats and violence of action that the troops face daily. Generally, the troops come in first and create secure, safe havens for the command to operate, a buffer zone if you will. With this buffer zone, many times a commander loses—or should I say never develops—a "violence of action" mind-set necessary to be effective in the arena. They usually resort to a management role in leadership. They are insulated from the hard decisions of combat and many times lose touch with the reality of battle.

Solving this problem is easy. The key is to spend a longer amount of time working with the troops. Ideally, a commander should be brought up from the troop level to ensure he has a proven and developed combat mind-set. Effective combat leaders should require their troops to be aggressive in combat and hold them to it in training. They should oversee training and hold the personnel accountable by being there. When you can smell the gunpowder, you're getting close. Don't get in their way, but be close enough to ensure they have the right mind-set and aggressiveness needed to accomplish the mission. If they don't, it is up to you to instill it in them or fire a TL.

As with the TL, the tactical commander should be under the scrutiny of his bosses, who require a higher standard because the survival of the troops is riding on his action or inaction. This is where decision making should become critical. If a commander cannot make a hard decision in peacetime and in training, he will not be able to do so under the stress of combat. Although this should be addressed and remedied, it is too often overlooked. Many of our high-ranking leaders seek out and promote those who do not rock the boat or make waves. As discussed at the TL level, troops should ask the command-level leader how far they are going to go to bring them back. In simple terms, "How many people are you going to kill to get us home?" If the leader cannot look you in the eye and give you a firm response regarding this issue, beware—you're headed for trouble. These types of leaders are the ones that will crumble under the stress of combat and will try to cover up all their mistakes once the battle is over.

### END GOAL

We must be able to apply the appropriate degree of force and discrimination, demonstrating a complete businesslike attention to detail; and if necessary, we must be able to kill with ruthless efficiency.

## KEY POINTS

- Develop one combat mind-set for all situations.
- Develop and practice your layered offense.
- Practice fighting through and never giving up.
- Have confidence in your drills and tactics.

# 2

# INDIVIDUAL LEADERSHIP

*War is an ugly thing, but not the ugliest of things. The decayed and degraded state of moral and patriotic feeling that thinks that nothing is worth war is much worse. The person who has nothing for which he is willing to fight, nothing which is more important than his own personal safety, is a miserable creature and has no chance of being free unless made and kept so by the exertions of better men than himself.*

—John Stuart Mill

- **WHY LEAD?**
- **ATTAINING COMBAT EFFICIENCY**
- **ACCEPTING RESPONSIBILITY FOR YOUR LIFE**
- **PERSONAL EFFORT**
- **BE SIMPLE AND EFFICIENT AND COMPETE WITH YOURSELF FIRST**
- **GET OUT OF YOUR COMFORT ZONE AND DEVELOP MENTAL TOUGHNESS**
- **TRADITIONAL AND NONTRADITIONAL APPROACHES TO ACHIEVING LEADERSHIP GOALS**
- **THE JOURNEY OR THE DESTINATION**

## SCENARIO: MOVEMENT TO THE CRASH SITE

We were on top of the target building, and AK-47 rounds and RPGs were starting to fly. We had just secured the target and rounded up twenty-two of the bad guys and were awaiting the convoys' arrival to ex-fil (ex-filtrate) them. We got the call that a Black Hawk helicopter had been shot down hundred to five hundred meters away from our position.

At the time, I did not realize it, but I had watched the combat search and rescue (CSAR) force fast-rope on to the crash site. As they were fast-roping in, an RPG hit the tail boom with a bang.

I watched the bird shudder and waver a bit while shards and strips of steel shot out of the area near its tail boom. The pilot did a heroic job of holding the bird in place while all the members of the team made it to the ground. Panicking and lifting off would mean almost-certain death for those still sliding down the rope.

The assault force commander (AFC) passed the word to the TLs that we would be moving on foot. The movement order was Rangers in front, us in the middle, and Rangers in the rear. The volume of enemy fire had increased, and I thought to myself, *This is going to be fun*. I briefed my team (Tony, Scott, Tim, and me), and we stacked in the courtyard by the interior of the gate. When our lead teams flowed out of the gate, we moved, staying on the left side of the street.

Initially, most of the fire was coming from straight ahead. As we approached the first intersection, you could strain your eyes and see the AK rounds sparking and glancing off the walls across the street, traveling from left to right. We would post our near side cover and wait a few seconds for a lull in the firefight. Generally a soldier with an AK will have to reload after twenty to thirty seconds of sustained semiautomatic fire. This is the time to take back the street and get your guys across.

When the lull took place, the cover man would roll out on his own, or he would get tapped out. Once he started his suppressive fire, we would dash across and pick a cover position on the other side. Generally, I would start my run not at the corner but several feet back so that when I hit the intersection, I was at a dead run while in the kill zone. I wanted to cut my exposure time and not give the bad guy the chance to get on his gun and track me. Wearing a heavy Kevlar vest and loaded with ammunition, you would need the extra few paces to hit full speed. Also, it would be a bit easier on your joints and ligaments to get the mass moving. In my case, it was like getting a heavy diesel truck moving. A diesel truck may not be fast out of the gate, but it can cruise at high speeds for a long time. We leapfrogged the entire team across the kill zone. Occasionally, we would also work with another team if they needed our support.

We continued to take fire from the front, rounds popping and cracking as they passed far and near. A few steps ahead and to my left, a Ranger took a hit in the leg and crumpled to the ground. To my amazement, no less than four to five soldiers stopped to try to help him. I yelled at them to keep moving and to let the medics handle it. My rationale was that if he got shot there, we could get shot there, hence it would be better to move forward and neutralize whoever was shooting. This way we would all be safe, and the casualty could be treated in relative safety.

Adding to the stress, the formation became a stop-and-go procession, reminding me of an accordion. As the firefight intensified, the lead elements were having a difficult time maintaining their momentum. Being in the middle and the back of the formation sucked, as we could not suppress to our front because of the friendlies up there. We had to move right or left and hope to avoid one of the incoming rounds that had missed everyone up front. Some of the Special Ops boys got pissed and moved to the middle of the street to lay down some fire to the front. Evidently they felt the lead folks were not putting enough rounds out.

We looked at the street we were on and saw that it had begun to narrow at one point. I took my team to the right side of the street and got on one knee to look over the three-foot wall of a house fifteen yards away. In the driveway, lying on his stomach next to a car was a Somali, moaning, a thick trail of blood flowing away from his chest. I thought about shooting him again, but I knew he was out of action, and I did not want to waste my ammo. I was worried more about the house fifteen yards on the other side of him, a perfect place for a shooter to come out and engage us.

Glancing to my left across the street, I thought, *I wish these dumb motherfuckers would get moving*. I watched a little bird swoop in extraordinarily low from right to left at about thirty feet off a rooftop and throw a green smoke grenade down the street to my left, about two blocks away. I knew that the crash site was in that area.

At this point, the formation was a bogged-down column that had lost its momentum. I shouted to someone to tell the AFC to follow me and curtly told the team, "On me," and began our movement down the left side of the intersecting street toward the site. We moved from cover to cover as best we could. Sometimes we would find a wall or outcropping to hide behind, but they were few and far between. When needed, we kicked in a door, cleared the initial rooms of the building, to create a safe haven out of the line of fire and to catch our breath and hastily plan our next move.

We did just that about thirty-five meters up the left side of the street. Once the initial clear was finished, Kim told me he had been hit. I pulled my tanto-style knife out of its Kydex sheath and plucked the battle dress uniform (BDU) jacket away from his skin and carefully sliced his uniform open in the area of his left kidney. I found the wound, and it looked like a high-speed entrance wound from a pencil-eraser-sized piece of shrapnel.

It had swelled up about the diameter of a golf ball and did not bleed. I told him it was not bad and he prepared for our next move. He immediately got back into the stack and got his mind on the game.

At this point, we were pushing forward into a hail of rifle, machine gun, and RPG fire. AK rounds were popping through the air, their sound similar to that of dry sticks being snapped. Occasionally we would catch the dust "poof" or "spark" out of the corner of our eyes when one impacted close to us. RPGs were coming every twenty to thirty seconds, and they would rock our world. We could see the orange glow of the rocket motor pushing the drifting dark mass toward us. A sharp explosion and a deafening impact would take our breath away. If we saw or heard it coming, we would try to get to a piece of cover and face away from it so as not to catch any shrapnel or secondary debris in the face or eyes. This could be a showstopper and put anyone out of action. As we continued to push forward, we caught movement to our right.

Looking across the street to our left, we saw two Somali gunmen running across an intersecting alley from left to right about twenty-five yards away. A Ranger was lying on his belly, behind a small foot-high knoll of dirt at the base of the alley, occasionally grinding his head from left to right, as if to get better protection for his face from the flying bullets, dirt, and dust.

Tony screamed at me, "Tell him to shoot the motherfuckers!" I immediately raised my rifle from the low ready, shooting about three feet over the back of the Ranger. I tracked the second guy and popped him twice and then transitioned to the first guy and hit him once. The first guy started to crumble, and the second guy tried to push him out of the way in the direction they had traveled. I immediately screamed at him, but I doubt if he heard me.

This whole action took about ten seconds, and we turned our attention to our front and continued our movement. We hit the next major intersection with our cover drill and began to move forward down the next street, which was gradually sloping downhill. I remember looking across the street and seeing Norm and his team in the dirt, glancing sideways at us in amazement. A gunman had the angle on them from a two-story building about two blocks away and was tuning them up. This is how combat is. One spot will be safe, while twenty feet away your buddies are getting chewed up from an angle that you cannot effectively put fire on.

As we continued forward, halfway down that street on the left side, we could see Rangers in a position to our front, huddled around a car under a big leafy tree. As we approached it, we looked to the right and saw the crashed Hawk, upside down in the alley, tail facing us. Without hesitation or guidance, Tony and Scott immediately moved across the alley to our front and began reorienting the Rangers and their fire. Fire coming from the alley to the left was heavy and getting worse. You could see the rounds zipping off the far wall as they impacted and continued their journey toward the downed aircraft and its defenders.

I figured we were going to be there for a while and needed to get the force off the street and make ourselves a home. We needed to start expanding our hasty perimeter. I started pushing on a heavy metal gate on the left side of the street, just short of the alley. It did not want to give. I called for my breacher, and after one or two seconds, a thought flashed in my head, and I told him to disregard the task. Looking at the gate and the immediate area, I realized we would need a heavy explosive charge, and there was nowhere to go to get away from the blast and the shock wave. We would probably suffer a casualty or two with a charge this big. It was now getting crowded there, and I figured I would give it one more try with the shoulder.

I was joined by an air force CCT attachment, who started shouldering the left side of the gate while I pushed on the right. His side broke first, and we were in. The gate opened into a courtyard that would later serve as our CP and casualty collection point. We did a hasty clear of the courtyard and found what appeared to be the front door to our right. I told my air force comrade to follow me as I cleared the four-room house.

Little did I know that Pete and his team had pulled my air force battle buddy out of our two-person stack. I entered a short central hallway and hastility cleared it. I spied the next room to my left and caught movement. With my nonfiring hand, I motioned for the family to come out to me. They were huddled at the back of the room. Cautiously, they came out toward me. I scanned them, and they appeared to be a simple Somali family that happened to be home when we decided to use their house as a firing platform and future CP. We secured them and placed them in an interior room of the house, safe and away from outside walls that were subject to rifle and RPG fire. I never saw them after that, and I assumed they stayed there the entire night.

## AFTER-ACTION COMMENTS

### *SUSTAIN*

- Ensure that worst-case movement scenarios are practiced and rehearsed.
- Team leaders control teams to ensure a fluid movement is achieved.
- Avoid rigid movement formations.

### *IMPROVE*

- When using combined forces, require all assets to practice movement techniques as a group not as individual elements.
- Ensure that medical procedures address casualty movement, treatment, and priority of care.

## WHY LEAD?

Albert Einstein was once quoted in *The New York Times* as saying, "Only a life lived for others is a life worthwhile." Each of us needs to find in ourselves that spark that drives us and ensure it is for honorable reasons. For without that belief and spark, the stress and chaos of combat will strip away the shallow armor of the self-serving leader and ensure his failure. The stronger the belief, the stronger the leader. In today's society, individuals seek leadership positions to enhance their social status, to gain power, or simply to please their parents. These reasons are insufficient to sustain the leader when the bullets are thick and soldiers are dying. One must truly believe in one's cause and one's fellow men to help ensure their survival. It was E. B. Sledge who said,

> *Until the millennium arrives and countries cease trying to enslave others, it will be necessary to accept one's responsibilities and to be willing to make sacrifices for one's country - as my comrades did. As the troops used to say, "If the country is good enough to live in, it's good enough to fight for." With privilege goes responsibility.*

Believe in your cause. The stronger your belief, the stronger your motivation and perseverance will be. You must know in your heart that it is a worthwhile cause and that you are fighting the good fight. Whether it is the need to contribute or the belief in a greater good—for your buddy, for the team, or for your country—find a reason that keeps your fire burning. You will need this fire when times get tough. It will help get you through when you are physically exhausted and mentally broken and you can only see far enough to take the next step.

On a personal note, I was rejected from the army when I was eighteen years old because I told them that I had asthma. I did not let that stop me. I worked for a short time in my second love—law enforcement. Longing for a career in the military, I had a local recruiter check the army database to see if my name still came up. It did not, and I enlisted, this time being a bit wiser on the question of asthma. *Probably just seasonal allergies*, I thought as I marked, "No" to answer the question.

I was fortunate to have the strength, stubbornness, and determination that allowed me to fight through or around the system. I knew deep down that I could do anything I put my mind to and that I would not allow minor setbacks or non-applicable administrative rules to chart my life. I refused to live based on someone else's decisions, or worse, indifference.

There are two things in life that you cannot control: Where you were born and who your parents are. Everything else we have influence over. Each of us has the ability to chart our own course. What plays an important part is how much sacrifice and hard work you are willing to do to get there. Most are not willing to go the distance, and instead they settle for what they are dealt with. Some have physical or mental handicaps, and I have compassion for these folks, but not pity.

I remember one of our guys got his leg blown off during combat operations, and yet he continued to serve for years in the army. He had a leg for running, swimming, walking; and he could downhill-ski better than anybody I knew. He was dealt a setback, but he attacked life and continued to serve as an exceptional soldier. He had an incredible attitude that focused on the goal and not on the minor obstacles in his path. This minor setback might slow him down, but not stop him.

I have also had the honor of serving with individuals who trained and inspired me to levels of excellence that I never imagined. Currently our society tends to churn out individuals that tend to ask the system, "What are you going to give or do for me?" We see this attitude all around us. Self-serving individuals concerned with their personal comfort and welfare beyond the norm. These individuals expect the system to take care of them at all costs. When I run across one of these individuals, it makes me want to puke. This attitude is damn near a form of communism.

I was fortunate to serve with men who wanted to know what they could contribute to their country and their team. They were forced to look deep, and they looked even deeper. They became accustomed to asking themselves every day, "What have I done to help the team or my country today?" It was a weight we carried daily. Max, a.k.a. CSM Maxim, would routinely give a short speech oozing with so much heart and emotion that it left you awestruck. He would put into words what you had to feel and believe in to be there. Wrapping ethic, integrity, honor, candor, duty into a short lecture, he would make you forget about your minor discomfort and feel like a small but important part of a machine that served a greater good, a "national asset," as he would refer to the unit.

Failure was not an option during my time in Special Operations. You took on a mind-set of mission accomplishment at all costs. Our equipment was the best that money could buy. Our support personnel and their equipment were on the cutting edge. If something was to fail, it would be the human spirit. We would continually push, hone, sharpen, and test our skills because we knew what was at stake. The survival of our country, our families, and each other rested on our ability to penetrate harm's way on demand and accomplish the required mission and return to do it again.

I learned early on that giving your best one time is not good enough. You need to be consistently good. I would not care to go to combat with an individual that was not consistent, good one day and not the next. I prefer someone whom you could count on the majority of the time. Yes, occasionally you have a bad day; but for the most part, you achieved your goals. If you failed a particular task, you would rehearse it until you got it right. This is what in my mind separates the average or mediocre from the successful. One-time wonders, those who could do it once and not *on demand* would fall by the wayside. The mind-set required a hunger for perfection, or at least near perfection. Occasionally you will find a natural. Most often you would find an individual who had developed a personal system in his life that was proactive and mentally prepared him to do the job.

## ATTAINING COMBAT EFFICIENCY

I firmly believe that we had the right people at the right time with the right training and combat mind-set instilled in them to ensure our success in previously mentioned operations. I have read several armchair quarterback versions that describe the mission as a failure or debacle. I disagree. At the operator level, we went in with our heads up, eyes open; we were lightly armed and punched through the best the enemy had. We stayed as long as we needed to against overwhelming odds and accomplished all our missions that day. First, we secured our intended target in a rapid and efficient manner capturing everyone alive. When a Black Hawk helicopter was shot down, we moved on foot through heavy fire and secured it until we were able to recover the bodies of our comrades that were trapped in the wreckage. While waiting, we turned the area into a meat-grinder, killing all hostiles that attempted to enter our perimeter. When we did decide to move, we brought all our dead and wounded with us and were ready to fight again hours after this fourteen-hour battle. Enough said, you get the point.

Politicians and military leaders will continue to put our forces in dangerous situations, and it is up to the individual and the team to ensure their own survival. How do we prepare ourselves for this type of intense and fast action? It begins with selecting individuals who have accepted responsibility for their lives and walk the walk as performers in the arena.

You must learn to influence yourself before you can influence others. Leadership has been said to be the influencing of others to accomplish a mission. Before you can become a leader, you must first learn to influence and master the physical, emotional, technical, and, if you will, the spiritual side of yourself. Your personal system should be a simple, efficient, and streamlined process to get you to your goal with some time to spare. You see, life is time, and time is moving. We can't stop it, but we can learn to use, manage, and live with it.

When aspiring to be a leader, you must first aspire to be a follower, to do the best job at each level you find yourself, before you set your sights too high, too quickly. Failure to manage your own life will only result in a failure to manage others. Leadership requires time management and efficiency, a system that starts with you. We must move forward on a direct and positive course. Without this course, we will either lose forward momentum or veer off and possibly end up somewhere we were not expecting.

Special Operations selection is a young man's game, requiring physical and mental stamina. Whether the selection process is geared toward the individual or the team player, one must be in the zone, so to speak, to be successful. You must be physically and mentally right and should be injury free, or not nursing an injury. Silly injuries in life can stop your career before it gets started.

## ACCEPTING RESPONSIBILITY FOR YOUR LIFE

---

Subject: Points from *"Dumbing Down Our Kids: Why American Children*

*Feel Good About Themselves But Can't Read, Write or Add,"* by Charles J. Sykes. He talks about how feel-good, politically correct teaching created a generation of kids with no concept of reality and how this concept set them up for failure in the real world.

Rule 1: Life is not fair. Get used to it. The average teenager uses the phrase "It's not fair" 8.6 times a day. You got it from your parents, who said it so often you decided they must be the most idealistic generation ever. When they started hearing it from their own kids, they realized Rule No. 1.

Rule 2: The real world won't care as much about your self-esteem as much as your school does. It'll expect you to accomplish something before you feel good about yourself. This may come as a shock. Usually, when inflated self-esteem meets reality, kids complain that it's not fair.

Rule 3: Sorry, you won't make $40,000 a year right out of high school. And you won't be a vice president or have a car phone either. You may even have to wear a uniform that doesn't have a Gap label.

Rule 4: If you think your teacher is tough, wait 'til you get a boss. He doesn't have tenure, so he tends to be a bit edgier. When you screw up, he's not going to ask you how you feel about it.

Rule 5: Flipping burgers is not beneath your dignity. Your grandparents had a different word for burger flipping. They called it opportunity. They weren't embarrassed making minimum wage either. They would have been embarrassed to sit around talking about Kurt Cobain all weekend.

Rule 6: It's not your parents' fault. If you screw up, you are responsible. This is the flip side of "It's my life," and "You're not the boss of me," and other eloquent proclamations of your generation. When you turn 18, it's on your dime. Don't whine about it, or you'll sound like a baby boomer.

Rule 7: Before you were born your parents weren't as boring as they are now. They got that way paying your bills, cleaning up your room and listening to you tell them how idealistic you are. And by the way, before you save the rain forest from the blood-sucking parasites of your parents' generation, try delousing the closet in your bedroom.

Rule 8: Your school may have done away with winners and losers. Life hasn't. In some schools, they'll give you as many times as you want to get the right answer. Failing grades have been abolished and class valedictorians scrapped, lest anyone's feelings be hurt. Effort is as important as results. This, of course, bears not the slightest resemblance to anything in real life.

Rule 9: Life is not divided into semesters, and you don't get summers off. Not even Easter break. They expect you to show up every day. For eight hours. And you don't get a new life every ten weeks. It just goes on and on. While we're at it, very few jobs are interested in fostering your self-expression or helping you find yourself. Fewer still lead to self-realization.

Rule 10: Television is not real life. Your life is not a sitcom. Your problems will not all be solved in thirty minutes, minus time for commercials. In real life, people actually have to leave the coffee shop to go to jobs. Your friends will not be as perky or pliable as Jennifer Aniston.

Rule 11: Be nice to nerds. You may end up working for them. We all could.

Rule 12: Smoking does not make you look cool. It makes you look moronic. Next time you're out cruising, watch an 11-year-old with a butt in his mouth. That's what you look like to anyone over 20. Ditto for "expressing yourself" with purple hair and/or pierced body parts.

Rule 13: You are not immortal. If you are under the impression that living fast, dying young and leaving a beautiful corpse is romantic, you obviously haven't seen one of your peers at room temperature lately.

Rule 14: Enjoy this while you can. Sure parents are a pain, school's a bother, and life is depressing. But someday you'll realize how wonderful it was to be a kid. Maybe you should start now. You're welcome.

---

Realize right now that you are the only one in control of your life. If you choose to make change, excel, or just be happy, it is ultimately up to you. You must believe this and put it into action to make it happen. Too many times, individuals quit before they get started, and then blame others for their misfortune. "I can't do this because..." They blame others for an accident, not getting promoted, and a host of other self-perceived problems. The bottom line: Don't sell yourself short on your ability to change your own world.

Life is like a compass or orienteering course. You can move fast and get somewhere, but it may not be where you intended. The skilled navigator will study his map, look at the terrain, and then choose between the fast and the slow routes to determine which is best for their journey. Many times in a difficult terrain, it is wise to pick out checkpoints to ensure you're on the right track. Life is the same way.

Occasionally you need to raise your head up and look around to see where you are and whether you are making the progress you expected. While navigating a multi-kilometer leg of a compass course in the Northeast, I had the choice of running the ridges, going up and down through generally light brush with a narrow footpath to guide me, or getting down next to the creek and hope the brush was not too thick and that I might find a small game trail. After a bit of up-and-down-on-the-ridge route, I said to myself, "To hell with this," and dropped into the valley below, finding not a game trail but a level superhighway-sized road that paralleled the creek and took me smoothly to my next checkpoint.

Sometimes you just need to look around and see what progress you're making for the effort you're putting out. Focusing on the terrain ten feet in front of you and daydreaming at the wrong time will cause you to sometimes take the hard path, and even overshoot a checkpoint. Worst-case, you might veer off course, you may end up somewhere you did not intend to be, and you might not be able to get back to your intended destination. On the land navigation course, it will cost you time and mental and physical energy to get back. In life, it might not allow you to fulfill your dreams due to age, a family commitment, or something of a similar nature.

## PERSONAL EFFORT

An old friend told me that life is like the saying on a soda machine, NO DEPOSIT, NO RETURN. I have found this to be simple and true. Today's society tends to produce a person who requires instantaneous feedback, rewards, or results with minimal input on their part. I have found that most rewards, or accomplishments are not easily won or earned. We get out of life what we put into it.

As mentioned earlier, in the physical arena, it generally takes two to three thousand repetitions for muscle memory to refine or instill a desired movement or action. Some remember the old saying of "practice makes perfect," but the reality is that you must focus on every move to ensure that you're doing it correctly. "Perfect practice makes perfect." The next question you need to ask is, "Have I met my standard or society's standard today?"

Society has become "kinder and gentler," in an effort not to hurt people's feelings or make them feel inadequate when developing standards. Most standards, whether it be for the military or the average police department, are set low to accommodate the bottom-feeders of life who lack the personal pride, motivation, or determination to rise above the rest. Routinely, we put military personnel and law enforcement officers in harm's way despite their lack of firearms proficiency and despite their being in worse shape than the hostile elements they are going to encounter.

A recent example of this universal lowering of the bar can be found in the Federal Air Marshall Service and their shooting program and standards. Initially, the program was top-notch and upheld only the highest shooting standards. Since the influx of thousands of new Sky Marshals into the program, the standards have been drastically lowered. The designers of the initial standards knew the dangers of making a surgical shot on a tubular target, such as an airplane. They knew that accurate and rapid shots were needed to neutralize threats, some of whom are possibly running down the aisle toward the folks driving the bird. You would not want an officer to miss and send rounds into the cockpit. This could be a showstopper for everyone on board.

Consider your own standards when developing combat skills. Shooting, for example, is a perishable skill and a personal skill. I would not want to have a TL hovering over me to ensure I dry-fire and practice. I should dry-fire on my own, two to three times a week, because it is the right thing to do. Dry-firing will ensure that I make a surgical shot with minimal rounds fired, killing the bad guy. It will help me keep all my rounds in my desired target and possibly not allow the threat to kill one of my comrades or any innocent people.

During dry-fire or dry practice, as some describe it, you go through all the mechanics of shooting from your ready position to punching the gun out in an efficient manner, aligning the sights and then dropping the hammer without disturbing the sights. The standards I practice are mostly fired at seven yards, and you have to hit the kill zone in one second or less from a high-ready position. So when I practice, I practice at not one second but at .90 seconds by using a standard shot timer that alerts you when to shoot and generates a second beep at the desired time limit. By practicing to be faster, I am not just trying to meet the standard, I am trying to exceed it.

## BE SIMPLE AND EFFICIENT, AND COMPETE WITH YOURSELF FIRST

Keep your life as simple and efficient as possible. How? Learn with each stumble or failure. Failure is only an opportunity to excel. The perception in which your mind chooses to visualize your challenges in life will enable your successes or enhance your failures. Learn not to worry, but to channel worry, anxiety, or dread into positive action that will enable you to succeed.

Set your goals and determine the best course for achieving them. Go back to the land navigation analogy. Find useful vehicles that can help you achieve your goals, and this may include important factors such as time and money. Airline pilots are a prime example. Many use the military as a stepping-stone to their dreams of becoming an airline pilot. They let the army, the navy, or the air force to pay for all that expensive flight training and hours. They use the largest aviation organizations in the world to support their learning while also serving their country. It is not only smart but efficient.

Another simple example is that of succeeding on an army physical fitness test. Most people dread it and fail to adequately prepare for it. Many will run only two miles to get ready for a two-mile test. Minimums. We have become a society that strives for the minimum standard, and this is how we live our lives. Failure to put in the desired time and effort will reward you with only middle-of-the-road performance. As I noted earlier, once in a while you will find one natural in the crowd, not often, but occasionally. For the rest of us who have to struggle to succeed, I would not have it any other way. I find reward and satisfaction in working hard for something and attaining even a small goal.

But back to the preparation for the physical training test. How could you more adequately prepare for a physical training test? First, get together with someone who is successful and ask them their secrets. Most often they are likely to tell you that you have to pay your dues by hard work and training. Making the time to train might be the key. You might have to sacrifice sleep and get up early to get workouts in.

Doing more to prepare will always help you. Set target goals and dates for your training. Use these as checkpoints on your personal map to see if you're heading in the right direction. Get up early on the day of the test and get warmed up by doing a warm-up run and exercise. Get your blood flowing and get your muscles right. Too many individuals are still tired and tight from sleeping, and their muscles now need to be warmed up enough to get them efficiently through the event.

Moreover, as an instructor, I see many students not taking the time to develop their own personal system, watching instead the other individual. Too many times we look at others with envy for a perceived talent or skill and fail to see the hard work and sacrifice they put into their success.

Train at your own speed and level of condition; draw a mental curtain to block out useless information and focus on what is best for you. If you try to go too fast, you have the increased chance of getting hurt, and then you'll waste a great deal of time in rehab, and you'll lose valuable time you could be using to move toward your goal. In addition, your strengths and weaknesses are different than someone else's. You must tailor your training to your individual needs.

A. It allows you to challenge your soul.
B. It teaches you the importance of teamwork.
C. It provides a mirror reflecting who you are.
D. It exposes all the good and bad in yourself.
E. There's no way to hide on a road march.
F. It strengthens trust in your leaders.
G. It toughens you mentally.
H. It beats complaining right out of you.
I. It orients you to authority.
J. It makes you think about others.
K. It matures you.
L. It makes you more objective.
M. It provides a frame of reference for suffering.

The road march is the crucible on which the soul is refined. Pulling a trigger is easy. Humping the load over the distance is where you find out who will be on the ambush site to pull the trigger on you. The road march defines you. Never quit. Come in ugly if you have to, but come in.

Don't be afraid to record a video of yourself and use it to critique the real you. We routinely have a perception as to what we look like during an action. The video camera will not lie. This is where perception and reality crash together. Once you get past the initial shock of what you look like, which we all go through, you can then methodically break down your actions and figure out positive solutions to your problems. You learn to accept the visual as what it is—real time, real truth.

## GET OUT OF YOUR COMFORT ZONE AND DEVELOP MENTAL TOUGHNESS

At the individual level, you need to get out of your comfort zone and push your limits. Too many times we fail to break out and change or experience the positive aspects of change. I have found that for a great deal of people, human nature is to seek the status quo. The military and the government usually set the bar low to accommodate the current social status quo. Whether to appease a social group or avoid hurting feelings or just make waves in the EO arena, we lower the overall strength and capability of our group when we lower the standards.

I use the dreaded military road march to illustrate a way of pushing through those walls of discomfort that we routinely hit in life. These are the walls that we need to go over or around to attain personal goals or objectives. Most individuals who have served in our armed forces can remember the pain of a road march. For infantry units, the road march is near and dear to their hearts. This reflection of the road march was sent to me and it goes as follows:

The road march is a level playing field that can enhance your physical and mental skills and endurance. The distance does not change for any race, age, or gender. The weight you carry and move from point A to B is the same for everyone. The variable in the equation is your mind-set, heart, and physical condition. The casual observer will look at a road march as a simple event, not requiring much time or preparation. This is far from reality.

Important factors that ensure success for this event are your mental and physical planning, preparation, and attention to detail. Failure to properly condition your boots and your feet are the surest tickets to failure. Proper training tells you how to buy and size the proper fitting boots. You learn to rotate your boots during training, breaking them in evenly. You learn to keep your toenails trimmed, or you will lose them through continual pounding and pressure on the front of your boot. Soaking your feet in Epsom salts several times a week will dry and toughen them. You choose socks that provide cushion and protection and discard worn ones. You pad your rucksack so that after three to four hours of constant wear, it has not punched a bleeding hole in your body that you can feel with every step.

Your training regimen is another learning point, starting with short distances carrying a lightweight load is the key to foot and physical conditioning. Starting with unrealistic distances and a heavy weight are two of the quickest ways to incur an injury. Your will only find torn-up feet, aching back and joints, and sores with going too heavy too fast. At this point, we also need to mention your water intake. Do not forget to hydrate properly prior to the march; dehydration can cause you to fail. Dressing too warm will cause you to overheat and give in.

Once you start the marches, you are going to hit a wall of discomfort. You need to know the difference between discomfort and pain and when to stop. If you condition yourself to stop when you feel discomfort and not pain, you're setting yourself up for failure. You're not only training your body to stop, but more importantly, you are training your mind to give in.

You need to plan, prepare, rehearse, and make it happen; then repeat this cycle over and over. By properly planning and rehearsing, you take away all the excuses, and you realize that you have all that is necessary to succeed or fail. Many Americans are weak-minded people who love excuses. As we used to say, the maximum effective range of an excuse is zero meters. You either succeeded or failed. Period. Find a physical challenge that requires two to three hours of mental endurance to be successful. You could start with a long walk, and then add weight. It could be a bike ride, or even a run. Whatever you decide to do, it will strengthen your mind, body, and spirit to a higher plane.

## TRADITIONAL AND NONTRADITIONAL APPROACHES TO ACHIEVING LEADERSHIP GOALS

Routinely we migrate through the process of life, from elementary and high school to college and then a job. This route provides the standard middle-of-the-road approach to learning and leadership. We may punch all the right tickets, but something is lost. That something is heart. Many individuals that I have watched endure this process don't have the internal fire, the killer drive or instinct for leadership. They are comfortable with status quo "management" instead of leadership and don't want to rock the boat with anything outside the proverbial box.

Traditional education does not usually promote aggressive and innovative leaders; rather, it produces a limited leader with a limited heart. It encourages students to rise to a standard that the system deems good or bad, but this standard is generally geared toward the masses and the status quo. Overall, our system lacks mechanisms for identifying those best suited for leadership and providing them a route to their goals and dreams. Sometimes leaders don't know what they don't know. Others simply don't understand what it takes, and they must be enlightened to fully understand that they can have an impact on those being led.

First, I didn't go into life wanting to be a leader. This may sound counterproductive. As I gained experience, I learned that I must have an attitude of trying to do the best job at the task at hand instead of constantly wanting to be promoted or aspiring to be a leader. I found that leadership challenges will eventually come. Some uninformed parents have serious problems with this idea. They feel that their little Johnny would make a natural leader. Most parents have not been around great leaders and would not know a good leader if they got smacked by one.

I suggest that students start by jotting down the good and bad points of leaders they respect and detest and use them as lessons or guidelines in their own leadership challenges. Mark down arrogance, excessive pride, and laziness as major counterproductive points to be avoided. Leave your ego behind on this journey and ratchet up the personal controlled aggressiveness. You can be aggressive and still be polite. You can be well mannered and not piss people off or be a threat to those above, below, or on your level.

## THE JOURNEY OR THE DESTINATION

What is more important to you? This may help answer the question of why you're doing what you do. Is it for prestige, fame, notoriety? Or is it to help your fellow man who is trying to make it through the same life, bombarded by the same struggles and challenges, trying to take it one day at a time?

I have come to the conclusion that *life is a work in progress, take one day at a time and enjoy each step, whether painful or delightful.* When life gets tough and it looks like the gators are going to get you, step back, take a deep breath, and start killing them one at a time, preferably the closest one first.

During your journey and challenges, you will need to keep your integrity intact. A Special Operations friend named Oz relayed a story of integrity and values to me. He said that life is like climbing a slow gradual hill with a rucksack full of bricks. The bricks represent integrity and values. As the hill tires you, some tend to drop a brick here and there to lighten their load. Many find that they reach the top of the hill with an empty rucksack. Some, though, keep all their bricks through the journey. Some who don't have all the bricks at the beginning of the journey pick one up here and there along their way. You may find that some steps might be unpleasant, but enjoy them all and try to pick a few bricks up along the way. Find that balance during your movement through life. Compare life to a navigation course and plan where you want to go and how you're going to get there. Don't be afraid to take an unorthodox path to get to your destination.

## KEY POINTS

- Establish personal reasons for becoming a leader.
- Ensure that you are willing to travel the necessary path to reach your goal.
- Compete with yourself first.
- Tailor your training program to meet your specific needs.
- Get out of your comfort zone.
- Enjoy the journey as the destination is too short.

# 3

# SELECTION

*Selection is a never ending process.*

—Old Special Ops saying

- **SELECTION TARGET—SURPASSING THE STANDARD**
- **INITIAL SELECTION FOCUS—TEAM PLAYER VERSUS INDIVIDUAL**
- **REALISTIC TESTS AND EVALUATIONS**
- **MENTAL PREPARATION**
- **PHYSICAL PREPARATION**
- **LONG-TERM FOCUS AND ONGOING SELECTION**
- **SETTING UP AND VALIDATING A SELECTION PROCESS**
- **SAMPLE GUT CHECK**

---

## SCENARIO: TARGET AK

We received word from an unreliable informant that there was a high-level meeting taking place at the target designated as AK and that there was a possibility of the top dog being there. Training our eyes on the target, we saw that there were guards present outside the building, which was a good indicator that something was amiss. The TLs reported to the tactical operations center (TOC) and began their planning session. The target consisted of seven to eight neat rows of apartment buildings, occasionally separated by a small house or building between them with an average of fifty yards of space between the structures.

Our pinpoint target was a small house in between the third and fourth sets of buildings. I thought it should not be too hard to find, but I drew a sketch map just to be safe. I stuffed it into my vest where my "chicken plate" normally goes (a chicken plate is a ceramic plate and the part of a vest that actually stops rifle bullets). The plates are heavy and bulky, and I normally did not wear one on operations where mobility was a concern. I would rather be fast and agile like a jungle cat than slow and easy to hit like a tank.

We planned the hit and moved to the birds with the pilots for our final briefing. The flight went well, and we were inbound when the gun birds gave the guard force an education in firepower about thirty seconds before we landed. I watched as the tracers impacted their target area and bounced up off the building into the sky. Our ride was smooth, and we were supposed to land directly adjacent to the target building. On our approach, Kim serviced the guard shack with a 40mm round just to keep any potential occupants honest. As with most missions, nothing is perfect, and our pilot's sharp eyes saved us again from running into a single power line that ran across the street in our landing spot.

We landed long and off our mark. We moved straight to a low wall that surrounded the compound and started looking for our target building. At night, under NVGs, looking down the row of buildings, it was difficult to see which one was one, two, three, etc. I was looking for a structure of sorts, but I could not pinpoint it.

We headed toward the first building and linked up with a sister team and started clearing apartments. They went downstairs, and we moved upstairs, leapfrogging with our security element. We charged the first door and began our clearing. The rooms were sparse—some furniture in the living rooms, maybe a bed and upright closet in the bedrooms.

We found a door that was locked, and I told Scott to charge it. As he did, a woman in the apartment came to me with a key, and I told her we did not need it. He called out, "Burning!" and we took cover and with a blast, the door was open. We cleared the room and found a case of local currency. It was only about U.S. $100. We left it and went outside, linking up with our sister team at the end of our row of apartment building.

I pulled my sketch map out of my vest pouch and looked it over. We were one to two buildings too far down. We moved down and spotted a house in between two of the large apartment buildings. As we moved toward it, we saw other teams ready to penetrate it, and joined them. They breached, and we entered behind them and began to flow through the target. Teams became mixed, and we would reorganize in rooms deeper inside the structure. Distraction devices were thrown, and their detonation would send up a dust cloud made up of all the filth and dirt on the walls and floors. Within one to two breaths, the back of your throat was coated with a nasty-tasting fine dust.

We kept clearing, and while I was following my partner into one room, a child that appeared to be about six years old tried to run out past us and ran square into the magazine of my weapon. The sharp corner left a small gash in his forehead, and he ran crying back to his parents, holding his wound. This room contained a heap of junk and debris that we had to move around. The next room we cleared contained a pile of junk covered by a nasty old cloth tarp.

My partner covered while I yanked the tarp away, causing the rats to scramble around the mess. *Delightful*, I thought to myself. This portion of the structure was clear.

We moved outside and found a door to another part of the building. We breached it and found that it opened into a kitchen and to a dead-end pantry. It was clear. Before the pantry on the left was a half bookcase up against a wall with a full-length curtain behind it. Pulling the bookcase away, we found a door behind the curtain. With a quick pull, the curtain came off, as did the entire door.

We entered the room and found a family lined up, facing our entry point. The kids were at one end, the father on the other, and he was holding out an infant as a shield. I moved and cleared left, and we secured the scene. I became pissed and told the boys to take all males over sixteen years of age and flex-tie them so we could take them back to the rear; I also told them to include the brave father who used his infant as a shield.

Outside we linked up with another TL who requested two of my guys for additional clearing duties. I assigned two of my men to him and told them where to link back up with me. We cleared a couple more doors in the immediate area, established security, and consolidated our prisoners. While waiting for my guys, the commander got a report of another house-type structure at the far end of the apartment complex. My men returned, and we headed for the new target with a sister team. We passed through a security position, and I let them know where I was going, and that I was going to come back through them. As we approached the target building, we found that we were hitting it from the rear. I dropped off at a back gate, and the other teams moved to the front.

My breacher put a hefty charge on the center of the metal gate, where the two swinging sides joined. He called out, "Burning!" and we took cover on both sides, out of the blast radius. A few seconds later, the charge went off with a loud *boom*. We moved through the smoke, and I caught movement and heard and crying off to my right.

Shining my gun light, I found a goat on the ground, quivering, doing the kicking chicken. As far as I could tell, the goat was a pet of sorts and came up and sniffed the back of the gate opposite the charge. You can probably guess what happened next. The charge went off and blew the goat ass over teakettle into the dirt. We moved past the four-legged casualty and moved to the back door of the residence. The door was recessed in an alcove, and we charged it. Another heavy charge was placed, so I tucked into a corner a few feet away and watched as fiery debris flew past me.

Looking at the breach point, I saw that the homeowners had placed a dresser up against the door to block our entry. The dresser was now three to four inches tall and spread out down the hallway. We entered and caught sight of our sister team down the hall. We cleared to them and joined them on the final room. As before, we entered, and saw that the father had the family lined up in front of our entry point, holding out his infant as a shield. I thought to myself, *What a cowardly low-life mother fucker*. I told the boys to secure him as he would be going for a helicopter ride.

We took our prisoner and began to move back to our first target building. We passed through our security position without incident and headed for the link-up. We deposited our prisoner with the prisoner-handling teams and got the exfiltration order from the boss. It turned out the two knuckleheads that used the infants as shields were high-ranking officers in a bad guy's militia.

## AFTER-ACTION COMMENTS

### *SUSTAIN*

- Be prepared to land off target and know the immediate and surrounding area.

- Have maps ready and available for all personnel.
- Keep discrimination rules simple and up to a high standard.

*IMPROVE*

- Think about dust respirators in nasty environments.

## SELECTION TARGET—SURPASSING THE STANDARD

The first order of business when structuring the selection process is choosing a safe and fair system that is geared toward the mission/job criteria or toward the individual's ability to learn the job. Having discussed the individual traits that one should strive to develop in their lives in chapter 1, we now need to look at the individual we wish to select as a leader. Should all leaders be volunteers, or should you determine who goes through the selection process? Does their immediate supervisor need to recommend them for the selection process?

These questions boil down to the issue of filling your selection process. Success is generally found in individuals that wish to voluntarily participate in the selection process because they have the drive and desire to do so. Occasionally, you will have to urge someone who has become too comfortable in their comfort zone and has realized they have found a niche and enjoy an easy and comfortable life. This type of person who needs encouragement or urging on is generally an individual who lacks self-confidence.

While I don't believe that someone should develop their self-confidence at the expense of others, I feel that every so often you can see the potential in someone and all they need is the right counseling and encouragement to spark a desire to expand their potential. In the end, all you need to do is show them the door and give them the opportunity to walk through.

## INITIAL SELECTION FOCUS—TEAM PLAYER VERSUS INDIVIDUAL

The selection focus—or I should say, the individual focus, is next. It is important to understand first that some team players will not do well without the team surrounding them for support. Some selection processes are geared specifically to the team concept and understand that they may have a weak individual and the team may help them make it through the selection process. Generally, individuals who do well in this team-geared process are not as strong or initiative oriented as those who do well in the individual selection process. The group selection process allows certain individuals to season, learn, and mature within a group, should they choose to, and the group will aid them in passing the selection process.

The individual process, on the other hand, is looking for a resourceful individual that can work for extended periods without contact and can muster the drive and self-initiative from within and does not need the security of the group to succeed. The structure of your selection process can be tailored for each of these individuals.

The individual selection process can be geared toward individual initiative and drive, with challenges pitting individuals against individuals or time. Time might be one of the criteria in the selection process. How do we establish a time standard for an individual? Several ways exist, but one simple method is to have a group who has already passed the selection process run the event cold (while timed) and then add all the hours together and divide it by the numbers in the group. You can then add 10 percent to 20 percent to the time to compensate for the experience level of the group, and you now have a standard. This can be done for land navigation movements, shooting courses, or typing tests. It simply does not matter.

The group test can be accomplished the same way—by taking a group and having them run or "validate" the drill, taking the average times and adding 10 percent to 20 percent. Then with this new 10 percent to 20 percent standard, have the entire group run it again and see how they stack up against the new standard.

## REALISTIC TESTS AND EVALUATIONS

Special Forces employ a team concept for selection, but also challenges the soldier at the individual level. Land navigation is one medium that equally discriminates against all players and provides a level playing field for candidates. In navigation courses, the ground does not change, nor do the distances. The weight and the equipment are the same for all. The ability to think and properly select routes is one variable that students can practice and become proficient in an effort to minimize route selection mistakes.

How does one get better at land navigation? Simply by taking the initiative and practicing and walking on their own time. I know countless soldiers that spend their weekends with a map, a compass, and a rucksack, practicing navigating cross-country and on roads. While navigating, they develop their strength and endurance as well as their mental preparation.

One will also learn the common mistakes made in land navigation, such as not walking too far. You see, as you walk, you become fatigued. Your pace will slow, and you will not travel as fast or as far. Your fatigue and your brain will say to your body that you should be there by now, and you are actually short of your target. Generally, the unskilled navigator will not trust the checkpoints they have established on their maps and will stop short and look for the point, wasting valuable time on their route. By getting used to the discomfort of the rucksack and the time it takes to travel at a certain pace, you will strengthen your mind and your body. Further, this not only shows initiative, it also shows heart. Read back to the road march quote, and it will put this in perspective. Total focus is required even if you're distracted by discomfort.

## MENTAL PREPARATION

Mental preparation is a personal thing, and as an individual, you will develop your own system of doing business. Using the land navigation analogy, some individuals will look at it as fun and program routes in areas that will not only challenge them, but will also offer them some visual stimulation. Some might take to the mountains and walk challenging routes on rugged terrain, while others might take to the beach and let the soft sand break down and build up their muscles while watching the surf. Some turn the walk into fun and bring along family members. Others see it as a challenge and push themselves to beat the time and distance goals they have set. Still others see it is a means to attaining a goal and simply "gut out" the walks. If possible, make it a fun and enjoyable ride to your goal. If nothing else, use it as meditation time. I remember going for hours on rucksack marches, and the scenery allowed me to clear my mind for hours, and I came off it physically pumped and mentally energized.

## PHYSICAL PREPARATION

Physical preparation is again a personal area where the individual needs to maximize their training time to accomplish two or three things while getting their workout in. The first rule in the beginning of your successful training session is to start slow and light. You want to let your body adjust to the walking you're going to do, and it may take several sessions to do this. There are no shortcuts that I have found to walking with a ruck, other than to put it on and hump it. Having said this, select comfortable and cushioned socks and footwear and watch for hot spots. Hot spots will generally start to appear because of soft feet or ill-fitting socks or shoes. The walking process will begin to condition your feet to your weight, while you are conforming your feet to your shoes. It may take several weeks to work the aches, pains, and soreness out. I have found that if you start slow, you really will not suffer too badly.

An aspirin after the walk will help your body heal. As mentioned earlier, soaking your feet in warm-to-hot water with Epsom salt will make them feel a great deal better, and the salt will dry and toughen them. If you develop a blister or two, Epsom salt after you march is the hot ticket. It will physically help your feet, and it will mentally relax you.

Too far too fast is a recipe for disaster. Too much distance will break down your muscles, your feet, and your spirits without a positive return for your time and effort. Too much weight too quickly will tear down your muscles and joints and is the fastest way to end a training program. I have watched countless individuals tear a muscle or blow out a knee by training improperly, causing them a temporary (months) or a severe (years or more) setback. Taking care of your feet is the key. One knee surgery can take you out of your window of opportunity for a lifetime.

## LONG-TERM FOCUS AND ONGOING SELECTION

The quote “Selection is a never-ending process” is true. A selection process can be structured as a brief snapshot of the individual or it can be a more detailed picture, or even a short movie. Even more, it can be a miniseries or trilogy, where it lasts for a year or more. There are selection processes that last over a year, sometimes two, where the candidate is on probation for a period of time. I still consider this part of the selection process.

The importance of the position you are selecting individuals for generally dictates the length of the selection process. You can get a snapshot of your candidate through applications and the paper trail they generate, but we all know the inflated paper trails that are being produced don’t accurately reflect the real candidate. I have seen the school chasers, those individuals who paper their walls with certificates from all the schools they attended, but rarely can they perform. Most of these chasers seek the schools for promotion rather than to bring back the information and make their people or team better. There are those who do have a sincere interest in bettering themselves or the team. Occasionally, they will go to a school; but generally, these people are few and far between.

School chasers, on the other hand, will have stunning résumés but are functionally illiterate in their fields and are often socially inept. Much of the time, they put so much energy in getting certificates, they never do their job, or their entire focus is spent on getting the “slot,” that they never have a chance to develop their team. Further, if they are always in school, they are never home to pass the knowledge on and develop their teams.

I understand if there are time and money constraints in your selection process, but if the job is prone to life-and-death high-stress situations, maybe the selection process should reflect that. With proper selection, you will quickly weed out the paper chasers and hangers-on in short order and find the individuals with substance and character.

## SETTING UP AND VALIDATING A SELECTION PROCESS

How do you set up a selection process? It can be announced or unannounced, or it can be a surprise process that is geared toward their job. It can be made up of one test, several tests, or multiple different tests that check several different areas of interest.

For the tactical team member, the selection can be made up of multiple endurance events that require the individual to think when they become tired or fatigued. It can focus on their individual ability to think in high-stress situations and not self-destruct when discomfort and fatigue set in.

A twenty-four-hour test will tell you a great deal about an individual. Calling them in late at night, unannounced, and then having them perform several low-intensity physical events throughout the night, denying them sleep, would be a good start. Then begin the day with stress shooting and practical scenarios. You could end early that day or run it into the night. The bottom line is that the more time, effort, planning, and thought that you put into the process, the better the candidate you will select.

As a general rule, most people will become cranky with lack of sleep but hopefully will have the professionalism and maturity to control and deal with it. If they do not control their temper and emotions, they will let it out for all to see. This will be a red flag during the selection process that is simple to spot.

How do we spot these red flags? Easy. Team leaders and members should be running the selection process and should be assigned a section or time period of the event. This way, if you use four teams to run your process, you can get four individual assessments of the individuals from four different perspectives. You can even keep the problems from one evaluation team to the next, where the next team to receive them gets a true picture of their attitudes and performance. This technique also spreads out the work and allows all the team members to see all the potential candidates and vice versa.

Team members walking a road march, for example, and then conducting a shoot afterward will have the same team next to them the entire way and one or two members shooting with them to show them the standard. It is up to your imagination on how you set it up.

## SAMPLE GUT CHECK

I used to perform what I termed *gut checks* from time to time on ROTC cadets. The date was known for the check and not much else. I originally started the "gut check" to see who would be authorized to wear a black beret as part of the Ranger Challenge in ROTC. The black beret was originally worn only by U.S. Army Rangers, but in recent years, the army managers decided to waste a great deal of money by taking the Rangers' black beret and giving it to everyone and their brother. Snubbing the Rangers and their heritage, they forced this elite unit to adopt another beret color and now watch and cringe as soldiers from every walk of life wear their new berets as though they are pizza chefs. Enough venting and back to the story.

For my gut check, a bulletin would be posted on the board for those wishing to try out or qualify. The notice would give a date and time and where to be, with a prescribed list of items that were to be placed in their rucksack. On the day of the event, I would walk in at the precise time close the door behind me. If you were late, you would wait until the next semester to try out.

I would then give the cadets a mission statement that was brief and limited. They were randomly selected and placed on teams. A TL was randomly selected and given five minutes to brief and ready their teams for movement. The course would have been scouted and run several times in the months prior, quietly and discreetly so no one knew the routes, distances, or events except for those administering the event. Five minutes later, we would be moving.

Generally, it would be dark out, and we would walk. I liked to walk to warm the group up for greater things to come. I also liked to check them for injuries as we walked. You could pick out a limp a mile away and quickly pinpoint a potential problem. I would ensure that TLs would constantly monitor their people for injuries and that they would continue to drink water. Sometimes we would walk for two hours, depending on the group's pace. I set the pace, but it was my goal that everyone would finish, no matter how slow they were. The slower they walked, the longer they carried the weight on their backs. There would be a price to pay for taking their time.

While we walked, the fear of the unknown would be the first stress that would settle into their minds. They had to deal with it subtly and learn to control it within themselves. Their bodies would begin to ache, and after the first hour or so, they would become warmed up to a point that their bodies were ready for more.

During one event, we would stop three to four miles into the walk for a break, and each cadet would pick up a sandbag to load into their rucksack. This would play on their minds and body. During the break, TLs would check to ensure their team members drank water, and they would also check for injuries and then report back to me. I would then personally check to ensure that they also drank water and then ask if they were injured.

During the movement, if someone fell back more than one hundred yards, I would have a van follow the group to pick them up anytime. This was a safety precaution, and I would have the individual shuttled back or taken to another point to help run the course. We walked with our rucks for another two to three miles and then took another break. Team leaders were told to have everyone drink water and put their tennis shoes on, placing their boots in their rucks and their rucks on the truck. This took no longer than five to ten minutes and then we would be on a group run. The teams would stay together on this run, and we would begin slow and steady. We would run for five miles through the town back to the college to a point where our obstacle course began. The rucks would be waiting for us, and we would get the sandbags out.

Moving as a team, the members would have to take their sandbags through the course, and they would deposit them at a point that they did not know about. I needed to build a fighting position in the course, and this was the easiest way to get the sandbags there. As the teams headed out, their instructions would be to put the sandbags down and negotiate the obstacle and then recover the sandbags. The pace would be a fast walk, but the team could travel only as fast as their slowest person. Halfway through the course, the teams would drop their sandbags in a predesignated spot, and they would then begin a jog on the course.

Overall length of the course round trip was one and one-half to two miles. Upon completion of the obstacle course, the group would be directed to pick up their rucksacks and move to the college stadium. There they would take the standard Army Physical Training test consisting of push-ups for two minutes, a break, sit-ups for two minutes, and then a break. Finally, they would do a two-mile run for time. They would then be told that the gut check was now over and they could return to the classroom where they started.

During my time at the ROTC assignment, I ran the gut check or a variation of this event seven or eight times, generally once a semester. I remember students of all shapes and sizes that were proud of what they accomplished for the day. I showed them that they could move twelve to thirteen miles through various unknown physical and mental challenges and still pass an Army PT test. As a matter of fact, most did as well on the test that day as they did when they took it cold.

I think it is important for leaders to have gut checks to ensure that the force is meeting a combat standard and not the minimum army standard. Today the army has a grading system for their physical training test that would make your head spin. In an effort to "equalize" everyone, it breaks down categories into male and female and then into many age categories.

> *When you can do what I do, you can go where I go.*
> —Old Special Ops saying

This effort to make everyone feel good is beyond me. There should be one standard for those who are subject to combat conditions, a minimum standard. Why? The last time I checked, a machine gun, its ammo, or a 100-meter sprint under fire is not different for a male or a female, a young or an old soldier. Why should a command sergeant major in the army with over twenty years in have to do less physical training than a private who is going to the same combat theater? Why should a woman who does less on a physical fitness test get a higher score and rating on her yearly evaluation report? It makes no sense. All this inequality does is attempt to make the individual feel better about themselves for doing less. Further, it weakens the capability of the entire force while providing a place for discrimination and dissention to grow.

## KEY POINTS

- Design a selection process around current tasks, learning abilities, or initiatives.
- Develop a level playing field to conduct your selection process where all individuals are equal—this will take away excuses from weaker individuals.

# 4

# TEAM LEADERSHIP

*Of Every 100 Men:*
*10 Shouldn't Even be Here*
*80 are Nothing but Targets*
*9 Are Real Fighters . . .*
*We are Lucky to have them*
*They the Battle Make . . .*
*Ah but the one, one of them is a Warrior . . . and he will bring the others back.*

—Heraclitus, 500 years BC

- **TACTICAL TEAM LEADER–THE KEY TO CONSISTENT SUCCESS**
- **TEAM COMPOSITION**
- **TEAM LEADER SELECTION**
- **TEAM LEADER DUTIES**
- **THE ADMINISTRATIVE TEAM LEADER**
- **THE TEAM LEADER AS A TRAINER**
- **THE TACTICAL TEAM LEADER**
- **TEAM LEADER TIPS**

## SCENARIO: THE CALL

It was a quiet Sunday afternoon, and the day was passing slowly. Some passed the time with an early-morning run or a physical training session. Others read or did their laundry in mop buckets by the shower point. As the day progressed, we watched as the crowds gathered at the reviewing stand, an area routinely used by the local leadership to give speeches to the local citizens. The reviewing stands consisted of a two-story structure with a platform that had overhead cover from the sun. The platform faced a large street and open area that would hold literally thousands of local people who came to hear from their leaders.

Some of these leaders we wanted to talk to. Many were key leaders in the militia who we were having problems with. The problem was getting to them. The back of the grandstand area had several structures adjoining it where the dignitaries' vehicles could pull up, deposit their guest, and then stage a quick getaway. The other problem was the crowd. Thousands would gather to listen to the leaders speak. During the speeches, the crowd would be "seeded" with gunmen, some obvious and some covert.

Trying to apprehend a key leader during a speech would bring too much chaos and collateral damage to the mission. Besides contending with the gunmen, the crowd could go either way. They could run, or worse, fight and start a riot.

If the crowd had erupted, the only recourse would be to cut them down. By doing so, we would provide a great deal of political ammunition to the other side because their media was there to record all the events that occurred. As rallies go, there would probably be numerous photographers and reporters there to catch the blow-by-blow action that unfolded. Even though it was possible to neutralize the right security people, as soon as they were down, someone else would pick up their weapon and try to engage us.

Now there was a body on the ground, in civilian clothes, riddled with bullets and no weapon in sight. It would provide a great one-sided media story of how U.S. forces slaughtered innocent civilians. No, we would have to wait and try to find a better location.

As the day wore on, things started to pick up in the afternoon. Intel from a source (snitch/paid informant) began to come in referring to key staff members of the local militia. They were to have a meeting shortly at a local commander's house. The source was to give a prearranged signal next to the commander's location, which would start our deliberate planning cycle. Until he did this, there was not much we could do. As the source started getting closer to the target building, TLs were advised and they then got dressed and headed for TOC. This sent a nonverbal signal to the ATLs that something was afoot and to upgrade their relaxed state of dress to one for combat.

The signal was given, and we planned a hasty assault on this target. The structure was a multiple-story building. We did our best to find the front door to the target and adjust our plan and assets accordingly. We sketched out our plan and got the word to move to the birds. As we started our final brief with the pilots and the team, we got the word that the target had changed.

We moved back to the TOC and were briefed that the snitch was too scared to give the signal in front of the right building and did it instead about a block away because of the presence of overt security and gunmen on the ground. He was told to move next to the right building and give the signal. He did, and we began our second planning session. This time, the building was easy to see, but the main entrance was not. We had planned for it being in the front, but once on the ground, we found that it was not.

The command gave the word to go, and we moved to the birds, re-briefed, and climbed on the pods. We stood by until the entire package was ready and watched as the now-routine flight took formation. The lead birds lifted, and we could feel the engine rev and the blades start biting air as our platform lurched forward to take its spot in the armada. We flew out over the water and then made a right turn over the countryside and back over the city. As we picked up our heading, I started looking for checkpoints and began to orient myself with the route. I scanned below and ahead for threats and looked farther ahead for the target building, keying off the actions of the lead birds.

I got the one-finger minute signal from our pilot, and we unhooked our safety lines, knowing that we were inbound. We were planning on landing in pairs, two and two, but the area was too tight for the two birds, and the second bird drifted back and took our slot. Our pilot decided that we could not fit and pulled out of the formation and did a quick loop and brought us back around. It only took about a minute, and I found myself looking for targets and threats as we made our way around.

By this time, three birds had deposited their teams and were lifting off. This left us with plenty of room to squeeze in. I remember looking at the target building directly in front of me as we drifted to the ground and seeing that the building had two flat roofs with a parapet and open patios. I was worried that someone might pop out and fire at us as they had sufficient time to prepare for our presence. We exited the pods and headed for what we thought was the front door. As we closed in on our breach point, I had a local male in my way that did not seem too concerned that we were there. I scanned him for weapons and dropped him with a leg strike/sweep. I wanted to get into his personal space and assure him and others nearby that we were not screwing around.

As we closed in on the breach point we saw that it was an open door that led into a small warehouse of some sort. There were pallets of food and grain stacked up in several areas around the floor. One of our sister teams was already inside looking for the entrance to the residence that we were supposed to be hitting. It was not in there.

I took the team outside and looked to my left, and along the wall were one-room-deep shops to the corner, about twenty-five yards away. This was when I first noticed that the AK fire was starting to pick up. Someone was shooting at us, but they were air balls, too high to discern the sonic crack of the round. We moved past the shops and rounded the corner. We immediately found an open gate to our left and, glancing in, saw the entrance to the target building.

The steps went up about five feet from ground level. I took the lead and went up the steps and caught sight of a man from bird two, pulling rear security for the team as they cleared. I did not hear gunfire, which indicated to me that they were not in contact, and I asked if the upper floor had been cleared. He said, "No." I told my guys to go upstairs. We moved quickly up the stairs and into the beginning of a short hallway.

The point called for a banger, and we flashbanged the hallway and cleared the first room on the right. The room contained only a mattress. As Scott moved past a window, he caught a burst of squad automatic weapon (SAW) fire from a friendly blocking position about twenty-five yards east on the corner. Some of the rounds narrowly missed his head, and most impacted on the outside of the window on the side of the building. We then cleared a small patio to our left and then turned right, out onto a flat open roof that was surrounded by a three-foot concrete slotted wall, or parapet.

As we broke out, I caught sight of a fireball about two blocks to the north from behind a concrete latticework on the rooftop of a building that was the same height as ours. The fireball was produced by a gunman firing an AK at us as we came out of the roof door. Kim, who was behind me in the stack, later told me the gunman was hitting the doorway above our heads, probably because he was shooting full auto and his muzzle was rising. I broke right with Kim while Tony and Scott broke left and began engaging the gunman. I told the guys to stay down and not to break the visual plane of the parapet wall. I was more worried about friendly fire than I was about the bad guys.

As I got on a knee and rolled on my back into the parapet, an M60 machine gun from another position began chewing up the wall on the other side of where Kim and I were hunkered down. Evidently, the gunner had seen and heard Tony's and Scott's muzzle blast and thought it was bad guys shooting at them. Everyone hit the deck, and I could see Norm getting ready to break out onto the roof behind us through the same door we exited. I gave him a hand and flashed the arm signal to hold, and he could see that we were under fire.

I got on the command net and called the AC. I told him to tell the blocking positions to stop firing at us on the target building. Evidently, things were hot on the street, and the battle positions were taking fire. The perimeter commander had the bright idea during rehearsals and training to put first lieutenants in charge of the battle positions and made the more qualified platoon sergeants almost non-players. This may work when no one is shooting at you, but when the game is on, that is what squad and fire TLs are for, to control their elements. One lieutenant cannot control the visual input and physical output of fifteen different soldiers at one time. Only properly trained personnel can accomplish this mission.

We crouched and ran back into the target building and moved downstairs to see if the initial team needed any help. We went to work re-clearing all the rooms and helping them consolidate all the prisoners. We counted twenty-two people that were on the target, none offering any resistance. As we did our secondary clears, I found a computer and a monitor. The computer itself was too big to bring back, especially during a firefight, so I pumped shotgun rounds into the monitor and the box to destroy it. Once we had concluded our secondary clear of the bottom floor, the prisoners were consolidated into the courtyard by the team who had apprehended them. The commander said that vehicles were en route to pick us up and our prisoners.

I couldn't see squat from the courtyard and wanted to avoid the gaggle that was forming in the open area. I took the team up to the roof and helped with the blocking positions with the mounting fire that the teams were facing. We moved back upstairs and went into the first room that we had cleared and taken fire from. It had a window facing south and east, and we could maintain a good visual on a hotel across the street.

Tony and Scott took the south window while Kim and I took the east. Looking out the south window we saw about half a block down the street a corner with a tree in front of it. The blocking position below us then lit up the corner with rifle fire. I saw a man poke his head out and then retreat. I watched the man reappear in the courtyard, and he began yelling back and forth to a woman that appeared to be his wife. I could not see a weapon in his hand, and he might have left it at the corner where he had been shot at a few moments before. He soon disappeared, and I pointed to the corner and told Kim to hit it with a 40mm round. He was unsure of which area I was talking about, so I launched four to five tracer rounds at the wall next to the corner.

Kim picked up on my rounds and launched a high explosive round downrange. The round hit in the Y of the tree, exploding and blowing every leaf off. I thought I could hear the blocking position below us cheer, and the man never did poke his head out around the corner again. As I continued my scan to the left, I could see a helicopter hovering and men fast-roping to the ground.

## AFTER-ACTION COMMENTS

### *SUSTAIN*

- Maintain momentum once the assault has begun and keep clearing until you find the target.

- Be aware of both friendly and enemy fire.
- Do not "gaggle"; there is always something you and your team can contribute, even if it is only pulling security.

*IMPROVE*

- Conduct coordination and rehearsals with friendly units and ensure that discrimination and ROE are adhered to.

---

## TACTICAL TEAM LEADER—THE KEY TO CONSISTENT SUCCESS

I have found that the key to consistent success in small and large-scale operations rests on the shoulders of the tactical TL. The team's performance, training, motivation, and attitude are a direct reflection of the TL's drive and professionalism. When properly selected, trained, and resourced, they will ensure that the mission is accomplished in a swift and efficient manner. Fortunately for me during my TL time, I inherited a team of warriors, which made my job almost effortless.

Why the TL? S. L. A. Marshall put it best in his WWII account of Squad Leadership. In *Men Against Fire,* Marshall determined that the five-person team is the most effective span-of-control for the combat leader at ground level. To help solve the span-of-control problem, one must first understand the team size in a tactical environment.

In volatile, heavy combat and under high stress ("bombs are bursting, the bullets are flying, and people are dying"), four- to six-person teams are the typical maneuver element. This is what combat veterans of WWII, Korea, and Vietnam I have spoken to, come to agree on. Why? Again, we must train and prepare for the worst-case scenario, that being the noise and confusion of the battlefield. Marshall pointed out that the span of control decreases with the chaos of battle.

The veterans with whom I have spoken remark that when the bullets start flying, you can only effectively control and maneuver four to six. They said the platoon and squad combat elements routinely broke down into small groups firing and maneuvering on an enemy position. Much of the problem results from the auditory exclusion that sets in as a result of the shooting.

The noise of the battlefield is deafening. I can remember wearing ear plugs when going into combat to help me retain my hearing when things calmed down. You see, the firing will eventually cease, and the bad guys will try probing you by quietly sneaking in. If you have not protected your hearing, it makes it a bit easier for the enemy to close in on you. Because of your hearing loss, you are now only using 50 percent of your senses to scan with. The other 50 percent is your vision. While scanning is effective in locating enemy personnel, many times what keys you into looking in a particular direction is sound. This also applies to your men and getting their attention. They must be continually scanning to catch your hand and arm signals, or you must move to them and scream in an already-deafened ear. Taking away that very important sense is an unnecessary handicap. Especially with today's sophisticated electronic hearing amplification and protection devices that can now protect your hearing and amplify the important sounds on the battlefield.

Another problem is that men are taught to spread out when they are getting shot at: the general rule is three to five meters from each other. With them spreading out and you protecting yourself from incoming fire, you will be strained to see two people to your right and two to your left at all times. In the forest or jungle, vegetation will hamper your line of sight. In an urban environment, any ninety-degree turn a team member takes will block your vision of him and his status. Again, this is your worst-case scenario span of control that we must prepare for.

## TEAM COMPOSITION

The makeup of any tactical team needs to be simple and efficient and must complement the tactical element. The sample team structure that follows can apply to military or the law enforcement tactical team:

Team Leader (TL)
Assistant Team Leader (ATL)
Breacher
Medic
Less Lethal
Gumby

All members of the tactical team are *trigger pullers* first, meaning that they all carry a primary weapon (rifle) and a backup (pistol) in addition to their specialty gear. Some, for example, may carry extra equipment. A *breacher* carries a manual breaching tool such as a ram or Halligan as part of his team responsibility, they would still carry both a rifle and a pistol. Some will argue that this is too much equipment and that only a sidearm should be carried in the above case. I disagree. For military operations, I prefer the breacher to be a completely interchangeable part of the team, and this will require both weapon systems.

Sometimes a small range of missions may also require the officer or soldier to carry a pistol as a primary weapon. Even in this circumstance, I would require them to carry a second pistol as a backup should their primary pistol fail. This mind-set dovetails into the layered offense mind-set and the principle of always carrying a backup weapon. As for the men who want to carry pistols as primary entry weapons, they will argue that they can cover and do the job with a pistol, but I remain skeptical. A quick trip to the range by the TL will find this out whether this person is lazy and trying to get one over on him or if they can really do the job with their sidearm alone.

*Team leader* duties require practice and experience. A team member should have several years of seasoning in tactical operations before being selected to the TL position. It would be a good idea for the individual to first spend one to two years as an ATL to validate and polish any needed skills and learn all the aspects of the job.

ATLs will generally get the needed experience as they are generally in charge of the team during the TL's absence. Besides the temporary leadership roles, the ATL needs to know how to perform all the same duties of the TL to include tactical planning and leadership counseling. (Occasionally during training scenarios, it is wise to "kill off a TL" and require the ATL to take over and accomplish the mission.)

At the lower end of the team is the team *breacher*. The breacher is probably the most important person in the stack. If you can't get into your target, you can't have mission success. I think it is smart to have the newest member of the team become the team breacher for one to two years. Why? Everyone on the team needs to know how to breach should the breacher be hurt or killed during an operation. Anyone, including the TL, should be able to pick up a breaching tool and breach a door in an emergency situation. Also, if the team gets split up into a two- and three-person element, you can have a breaching-capable shooter with each group.

Routinely in the law enforcement arena, the biggest person on the team becomes the breacher, and they are saddled with the job until the next hulk arrives. This is primarily due to manual breaching being the primary breaching means in the law enforcement arena.

Next on the team are the specialty skills that team members need to possess in order to be a diversified asset. These skills can be *medical, less lethal,* or *communications*, the latter depending on how sophisticated your communications program is. As with breacher skills, medical skills are equally important. The first responder to a medical situation is generally going to be you. Whether you, your buddy, a hostage, or an innocent bystander is injured, you will be the first one to provide medical treatment. Everyone on the team should carry basic medical supplies and know how to use them.

The common response is to let the dedicated medic handle the injury. The reality is that the team medic or dedicated medic may not be able to get to the injured person because of the intensity of fire. The first priority should be to neutralize the threat and then care for the injured, otherwise more friendly forces will be at risk. One dead or injured body does not justify two. Once the bad guys are taken care of, the easiest way to stop the bleeding is to put direct pressure or a torniquet on the wound. Someone else will probably be cutting clothing, exposing the wound and getting bandages out. Pretty soon you will have two to three people working on one gunshot wound.

For ease and efficiency of treatment, everyone should have the same medical package and training. The team medic should be the subject matter expert for the team and may carry a few more bandages should the situation require them. In the law enforcement arena, it is just as critical to be medically trained and proficient because emergency medical service (EMS) personnel generally cannot come into a target until the entire area is secure. In the case of an active shooter scenario, such as the Columbine High School incident, you had scores of innocent victims that needed immediate treatment. I tell my LE classes that killing the bad guys is easy, but after it is accomplished is when the real work begins.

Individual officers are generally the first responders to an active shooter situation, and once they have security and have neutralized the threat, they should immediately revert to providing medical support. This is where the real life-saving work will begin. Rapid hands-on to prevent blood loss is the key.

Next, if there are any specialty missions that a team needs such as *less lethal, chemical,* or *shields*, then the team should have a dedicated expert to turn to. With the variety of equipment available, there is a great deal of information that needs to be tracked and passed on to team members to keep them on the cutting edge. Law enforcement teams are saddled with multiple layers of less lethal taking the form of handheld gas, gas launchers, tasers, and much more. Proper employment techniques require proper training and certification. This will also come into play when starting collective training with various teams.

Finally, sniper teams can maintain the same generic structure as the average assault team, but will add their special skills to the mix. I prefer to select snipers who have proven themselves as assaulters for at least two years. This seasoning allows the new sniper to understand the importance of reporting correct and accurate information to the command and assault elements. Having breached multiple barriers as an assaulter, the new sniper will know what information is critical when reporting breach points instead of passing a bunch of fluff or useless information.

Snipers will also know that when they perform their area recons, the important information includes the routes and obstacles that an assault team will encounter when approaching a target. Sniper teams have the ability to break down into to two sub–shooting teams for tactical operations in permissive environments such as your local communities. For the woods, I prefer to keep the sniper elements as a package, four to six personnel, so they can provide their own security and put up a better fight while in Indian country.

## TEAM LEADER SELECTION

As stated earlier, the TL should be selected from proven ATLs, those who have consistently and successfully led a team or part of a team in the TL's absence. ATLs should be able to run a team in the absence of a TL and should have the same qualities and attributes of their leader. If a leader is injured, killed, removed for cause, or it is just his time to rotate, the senior assistant should be selected to fill the vacancy.

Should the group consist of both sniper and assault teams, you might want to put an assaulter in an assault position and a sniper in a sniper TL role. Generally, these individuals are specialized by this time, and they will be more efficient in their specialty field. Occasionally, you can cross-pollinate the teams if the individual has the willingness to change over. While I was serving in Special Operations, it became the goal of every operator to stay in an "action guy" status as long as possible before rotating to an instructor's job and getting away from the action and door kicking. Once you become an instructor, you are known as an ex-action guy. As I saw my time approaching to become a TL, I realized that after two years of service in that position, I would be sent to an instructor's slot. So, I volunteered to do two years as a sniper. This would give me two more years in the trenches and the possibility of participating in a few more real world missions.

Being a sniper also gives you the best of both worlds. Having paid their dues as an assaulter for two years, they can do both sniping, tactical reconnaissance, or assault work as needed. It was literally the best job I had, next to that as a TL. As I rose to the ATL position on a sniper team, the TL of the assault team I grew up on was killed in a parachute accident. I was selected to return to this team as a TL.

It was a tough and solemn time for a while. I knew all of the men on the team, and they were exceptional individuals. The loss of Bubba, the team leader, was devastating. They allowed me back into his position, and I quietly took control of the team. I simply had to sit the guys down one day and let them know it was all right to grieve and that I understood. I also let them know that I was not Bubba and that I wanted to learn their system of doing business versus converting five people to my way of thinking. It is much easier that way. My job as a TL was to get the job accomplished as smartly and safely as possible. If they did not violate any security protocols with their tactics, I had no problem with how they accomplished the mission.

You see, in Special Ops, each group may have a slightly different way of doing things like tactics and planning. There is no problem with this, and they must be allowed their "creativity," if you will, to figure out the best way to get things done. A pitfall for TL selection is appointing TLs for anything other than their performance. This can be disastrous.

I have witnessed this firsthand in both the Special Operations and law enforcement community where friendship, gender, or ethnic appointments had life-endangering results. In Special Ops, we had one individual who had the rank but did not have the skill or the talent to do the job. The command appointed him to a TL position, and he ran the team into the ground. Finally, after an issue with a travel voucher, they sent him away from the unit to another special forces group, where he proceeded to screw up a special forces "A" Team. Instead of solving the problem, they flushed it to another area of the army for someone else to solve.

In the law enforcement arena, I am aware of an incident where an unqualified politically appointed TL pushed a hard hit on a barricaded person, where it was not needed. The bad person was justifiably neutralized, but in the process, a friendly officer was shot and permanently crippled. This is unacceptable. Out of this one incident, the team had already lost five individuals. This could have been easily avoided with the proper selection of a TL from within the ranks.

Too many times, organizations put restrictions on where TLs come from and appoint TLs from outside the tactical unit. Once in a while, it will work with the right individual, but generally it is a recipe for disaster. The team generally has to retrain the TL in tactics and techniques, wasting precious time and reinventing the wheel. Proven experience should have been a prerequisite for their initial appointment. In effect, the team who has to retrain a TL has to take two steps backward before they progress.

## TEAM LEADER DUTIES

Simply put, the success or failure of the team lies on the shoulders of the team leader. Their success is your success, their failure your failure. A TL's first job should be to get to know the team and assess their strengths and weaknesses. The getting-to-know part should be easy with personal counseling sessions.

The assessment part will depend on the training schedule and the variety of upcoming training. By simply conducting routine range fire, you can begin to evaluate their individual skills. Simply by doing a team run and physical training session, you will build morale, and it will give you an idea of their physical preparedness.

Once you're assigned the role, my suggestion is that you come in quiet and watch for a period of time. Again, I have found it easier to learn the team's SOPs than to try and change everything at once and start from scratch. If their drills are sound, why not stick with them? You will have time during your assignment as TL to slowly change things that you disagree with. Massive change from the start will only be met with mental resistance. Team members will either think that you have no faith in them, their tactics, or their techniques. If the tactics are unsafe or not sound, by all means change them. But don't change them to show them who is boss. This is an ego problem that you should have fixed at the individual level. Not only is the change silly, it is also counterproductive to the team morale and growth. Training time is a precious commodity and should not be wasted.

As a new TL, you should come in early and stay late on your own, reviewing the team's SOPs and checking out equipment, etc. You might quietly inspect the condition of the team's equipment, weapons, vehicles, etc., to get a better feel for their preparedness level. Ask questions of team members about their team specialty in a nonthreatening way, so that the team member you're talking to feels they are educating you rather than putting you on the spot. Also, by coming early and staying late, you get to see who is doing the same. You will get a better feel for the personnel on your team and their strengths and weaknesses.

## THE ADMINISTRATIVE TEAM LEADER

The administrative side of the TL can be as important as the combat or the training side. Up until the last couple of years, the military Special Ops were not killing a terrorist every day. Most of their time was spent on routine day-to-day training. Keeping up with volumes of paperwork the military and law enforcement require can be a constant chore. You owe it to yourself and your team to learn to type and to run basic computer software programs to ensure that you can efficiently handle the required paperwork. Most action guys will call this a sacrilege, but if you learn to become more efficient with training schedules, evaluations, awards, and after-action reports, the more time you will be able to spend on the range, with the team doing the important stuff like training and rehearsals.

You also assume a moral and professional obligation to ensure that your team is taken care of with regard to yearly evaluations and promotions.

As a Reserve Officers Training Corps (ROTC) instructor, I saw cadets commissioned that did not write a term paper in their four years of college. I was shocked. The problem that comes with the inability to write is that you cannot properly describe on paper with accurate detail how good or bad your team members are. This is especially important when your team members are competing for promotion against other officers or service members. Generally, they are not competing against each other, but rather, it's a question of how well their supervisors can project thoughts and ideas on paper.

I have seen numerous individuals in my career suffer because they had a supervisor who could not articulate a clear thought on their subordinates' evaluations. Evaluations suffered, as did individual awards. Those afraid of writing were also reluctant to put their people in for awards because of their poor writing habits. When they did put in awards, their narratives were generally too weak, or poorly written to support the recommendations. Awards were kicked back or downgraded—and who suffered? The individual.

As luck would have it, the unschooled TL would generally have a great boss that could write, and their evaluations would never suffer. The TL would carry on with their attitude and demeanor of non-education or self-improvement in this area, smugly getting promoted, thinking *I got mine*, or *The administrative stuff is not for an action guy like me*. This piss-poor attitude only trickled down and screwed over their subordinates.

## THE TEAM LEADER AS A TRAINER

Training for success in all assigned missions is the primary goal of the TL. Ensuring that team members are on the cutting edge will help ensure their survival and the survival of the team in a high-risk situation. Through your initial assessment of the team, you should have noted any deficiencies in their training or skills and prioritized which training is the most urgent.

Generally, training that will save them from death or great bodily injury will be the first priority. Many times something as simple as weapons or training safety will be a good start. Many leaders have found that as a result of a training accident, training of all types will come to a screeching halt. Although not as much in the military as in the law enforcement arena, an accidental shooting can halt effective training for years while all the fingers point and litigation runs its course.

It is of the utmost importance that the TL, the ATL, and the entire team constantly monitor safety, especially with units dealing with weapons and live-fire training. A simple system of checks and balances needs to be established to ensure that no one is hurt or injured. Just as important is setting the example for your team members to follow. The TL needs to set the example with safety, weapon handling, and discrimination. Once you develop your safety routine, you can then properly execute high-risk training. This system should extend and dovetail into flat-range fire, live-fire CQB—and then to combat with little or no changes.

Start by looking at the collective team skills and decide which ones the team needs to work on the most. Use all the assets available to ensure you maximize your training time and effort. What many leaders forget is that you do not have to be the subject matter expert in all areas, you simply have to manage them. For example, for internal team training, you can task out members of the team to set up, rehearse, and execute training for your element. If medical training is your priority, task the team medic to set up and deliver the training. Using this technique will spread out the responsibility and require each person to know the subject matter they are going to teach. If they fail to put out a good block of instruction, it will readily show in the training, and generally the peer pressure will ensure that they are not unprepared again. The TL also can counsel, face-to-face or in writing, the person responsible for the training.

A simple technique to begin this process is for the TL to plan, prepare, and execute the first block of training to set the standard for future team training. This will take away all the excuses from team members in future training when you get the occasional comment, stating that they did not know what was expected. The TL should ensure that all classes, presentations, demonstrations, and handouts are up to standard; and he should use this as a teaching point to ensure that the team understands what is expected of them in the future. And in the future, the TL should be briefed by the individual tasked to perform the training, to ensure it adequately covers the desired information. This will help maximize each and every training day.

As a final point in team leadership, the team leader should require team members to step forward at all times on their own and take initiative. Whether it be something as simple that needs to be done like cleaning up team equipment or someone rising to a tactical leadership challenge, operation personnel should understand that setting an example is the right thing to do. Members should not be penalized for stepping forward and taking initiative. They should be rewarded for it. Too many times, members are scolded or chastised for taking initiative. This will stifle initiative and aggressiveness and will result in a wait-and-see attitude in combat, which can be extremely counterproductive.

## THE TACTICAL TEAM LEADER

The transition from training leadership to tactical combat leadership should be transparent. *Live the example in training and mirror what you're going to do in combat.* Adhering to this philosophy will make your transition to the battlefield seamless. If you have built a realistic training system that mirrors combat and you train to exceed the standards, you will do well. The only change is that you may lose people. I understand this can happen in training when individuals are injured from time to time, but combat will induce more permanent losses.

As a TL, understand the concept of risk versus gamble. Look at your training, your battle drills, and combat with the understanding that there will always be some risk involved. Never gamble or just throw the dice and wish for a happy outcome.

Choose tactics that are simple and safe to execute and will ensure the greatest chance of survival for your men. We do a dangerous job, and the team needs to understand this. This is why we carry guns, wear layered bulletproof material and all the protective equipment we do. People will shoot at us and will try to do us harm.

As a TL, be smart about when you have to push the fight and when you should not push. We can be at the top of our game and still get killed from an un-aimed AK in the hands of a child two blocks away. When it is time to get after it, do so with all your heart and assets. Sometimes, though, step back and ask yourself whether this is the smartest course of action. Don't be afraid to get independent opinions from your men or other TLs.

As a final point, the TL is the first line of fighting leadership. You need to be technically and tactically proficient with all your weapon systems and set the standard for the newbie on the team. They should aspire to be like you one day. If you live the example, you set the stage for success. You also take away all the excuses, should anyone want to fall back on one.

## TEAM LEADER TIPS

Rotate TL planning responsibilities once you have become proficient and skilled at planning team and group assaults on targets. Begin by bringing your ATL with you to planning sessions and allowing him to see the process and expose them to what is going on. Say, every third or fourth planning session, start with your ATL and assign him the job of target planning for your team under your supervision. Once your ATL understands the basic planning concepts and requirements, start grooming another team member by allowing him to see how the process works. Some TLs see this knowledge as power and wish to keep their lower team members in the dark and ignorant of what goes on. Exposing them to the system is the best way to help develop subordinates into future leaders. In the near future, you may be losing an ATL who has been selected to take over a team. If you have a replacement in mind, start bringing them into the leadership planning cycle early and get them involved in the mission planning process.

As with the exposure to leadership planning, rotate team duties and responsibilities every two to three years to round out everyone on the team and to prevent burnout. As suggested before, rotate everyone first through the position of breacher and then give them a change of pace with either medical or less lethal. This will help prevent burnout and give the individuals a chance to refocus and excel in another area. Also, as a breacher, you routinely enter the room last and never get a shot at the number one or two position on entry.

Another positive point about bringing in the new team member as the breacher is that the TL has the chance to evaluate and assess the new officer for one to two years. Some organizations have a one-year probation time for new personnel, and this allows the TL to do a proper assessment. It also gives the new team member a job that is important, but not the immense pressure of running point. Running point requires a seasoned individual who is consistent and has been exposed to all the possible situations that one can run into. In effect, the point person is running the team for a short time in lieu of the TL.

For control purposes, find the best place to control your team. Some feel it is up front, but I suggest a place in the middle of the stack so as to see both the front and rear. The TL should ensure that all members of the team are trained to the highest level because on fast-moving tactical operations, contact can be made anywhere and by any member of the stack. For this reason, everyone from the point person to the rear guard needs to be dialed in and ready to do business.

Further, as a TL, I use a system where my team members get hands-on and either I cover or supervise. This is important because if I get caught getting hands-on, I am not doing my job as a leader. My job is to supervise my team and ensure that we maintain 360-degree security at all times. Wrapping someone up with flex ties and searching them does not allow me to do this. Sometimes you will find yourself in this position, but team members should be trained to see this, step in, and take over, freeing me up to do my job.

## KEY POINTS

- Team leadership means living the example for your team.
- Understand and apply the concept of *risk versus gamble.*
- Continually train your junior leaders to fill the next higher positions.

# 5

# ORGANIZATIONAL LEADERSHIP

*We trained hard... but it seemed that every time we were beginning to form up into teams we would be reorganized. I was to learn later in life that we tend to meet any new situation by-reorganizing; and what a wonderful method it can be for creating the illusion of progress while producing confusion, inefficiency and demoralization.*

—Peter Arbiter, Roman Legionnaire, 210 BC

- **DEFINITION OF AN ORGANIZATIONAL LEADER**
- **SELECTION OF THE ORGANIZATIONAL LEADER**
- **ROLE OF THE ORGANIZATIONAL LEADER**
- **TECHNIQUES OF LEADERSHIP**
- **PROBLEMS OF ORGANIZATIONAL LEADERSHIP**
- **WHEN TO SAY NO**
- **TEST, EVALUATION, AND INTEGRATION OF NEW EQUIPMENT**
- **ENSURING THE COMMUNICATION PROCESS IS WORKING BOTH WAYS**

## SCENARIO: BAD DOGS

Our next new target came to us via some disgruntled neighbors. While sitting on one target, waiting for another intel hit to come our way, some locals from the neighborhood came to our position and informed us that the people who lived next door to their residence were part of the bad guy's militia. We assembled our group, drew up a hasty plan, and headed out on foot to our new target. We moved two or three blocks, made a left-hand turn, and the target was the second house on the left.

Our team's mission was to move to the back of the residence and secure it. As was normal for the area, the house's yard was surrounded by a five-foot concrete wall. We did a hasty visual clear of the backyard over the wall. Being the number one man, I was then boosted over the wall. The T post from a clothesline was an arm's reach farther than I could manage, and I dropped onto a wooden chair that someone had placed against the inside of the wall. Upon touching the chair, my feet went through it, and now I had the ring of the seat around my legs. *Not a good start*, I thought.

I stepped out of the chair and took up a security position while the rest of the team maneuvered over the wall. Once we were in the yard, the fun began. Teams started hitting the front side of the house, and it awoke the three dogs in the backyard with us. The dogs decided they were going to do their job of protecting the yard and headed toward us, the new strangers in their world. As one got close to Gary, he fired off a 5.56 round close to the lead dog's head. The muzzle blast from the CAR-15 is wicked, and all three decided they did not want to play with the strangers. They ran up some back steps to the house and huddled together shivering, still wondering what the boom was all about.

We penetrated through the back door of the residence and linked up with the team on the inside. Coming back out, we ran into another team who was placing a charge on a locked door directly around the corner from the stairs and the dogs. The blast channel and overpressure was about to come around and give the dogs another dose of sensory overload. The team called out, "Burning!" and we all took cover, everyone except for the dogs. The charge went off, and the blast wave hit them, and that was all she wrote. They were last seen hauling ass south past a blocking position. The target turned out to be a dry hole. No guns, equipment, uniforms, or anything to indicate that bad guys lived there. As best we could figure, our informants were the neighbors that had a beef with those who lived in the house we just finished remodeling. We aim to please. Not a door was left intact, and the vehicles had a few more holes in them than when we started.

Simple, common-sense decisions by our commander allowed us to shed our heavy vests. We were able then to conduct numerous missions such as the one just described. We developed simple SOPs, such as keeping a one-quart canteen on our belts or in the cargo pocket of our pants and drinking it prior to a hit. Once the target was secure, we would refill the canteens and drink some more while pulling security and waiting for intel to designate a new target. This enabled us to conduct raids on demand without going in burned out or too tired to properly focus.

## AFTER-ACTION COMMENTS

### *SUSTAIN*

- Don't be afraid to make smart decisions using the risk versus gamble analogy.

### *IMPROVE*

- Be careful of old wooden chairs.

## DEFINITION OF AN ORGANIZATIONAL LEADER

*Organizational leader* is a generic term that I will use to describe a leader who is not in the first line of fighting leadership such as the TL, but is rather an administrative or tactical commander/leader who supports the force near or afar. Why make this distinction? Because the men at the tip of the spear see this distinction and see a difference. Typically, men in combat see organizational leaders in a different light. They each have different jobs to accomplish in peacetime and in combat. They also follow a different selection, training, and career path. Both the jobs of the TL and of the organizational leader are equally important in supporting the troops. That is why organizational leaders should work closely with TLs to accomplish the mission.

Who controls the planning, support, and equipment? It should be a combination of the tactical TL and the organizational leader. The organizational leader's role is not to dictate policy to those going through the door, but rather to help facilitate their actions.

## SELECTION OF THE ORGANIZATIONAL LEADER

Selection of the organizational leader should be similar to that of a TL. If you have a good system, apply it equally across the board. My first choice would be to promote from within and directly from the TL ranks. Many will argue that an organizational leader needs to be an officer that has gone through all the proper leadership gates to get there. I disagree. Leadership is leadership, and there should be one standard for both those enlisted and officers. There has always been a visible division and rift between NCOs and the officer ranks, and I doubt if it will ever change, until there is change in the selection and training system for officers. Ask any performer in the field who they would choose to lead them in a high-risk environment: the person with four years of college or the person with four years of experience? The answer is simple.

I do not see the entrenched system of leadership in the military changing in the near future, so we will probably have to work with what we have. If the organizational leader is to lead or command troops, then the selection criteria should be based on just that—experience and leadership in that chosen field. Military Special Operations has done a better job at selecting and integrating officer leadership than some of their counterparts in the regular army and in the law enforcement community.

Generally, Special Operations forces require their leadership to endure the same selection process as that for enlisted men. Once this is completed, they then branch out to their separate tracks in their chosen profession. An advantage the military has over law enforcement is the way soldiers are immersed in leadership from day one of their enlistment. Fire team and squad leaders are selected in basic training, as are platoon sergeants and platoon leaders. Leadership principles are pushed down to lower levels throughout the initial training process.

## ROLE OF THE ORGANIZATIONAL LEADER

Simply put, the role of the organizational leader is to best support the men in the arena. It is not to micromanage or overcontrol these men, as this will only affect your ability to move through Boyd's Loop during training and tactical operations. Responsibility, trust, and empowerment need to be pushed down to the fighters and TLs at all times to ensure mission success with the least number of casualties.

Span of control is important at the team level, but it is now mental control that needs to be discussed. Combat consists of TLs solving individual problems at their level to accomplish a desired collective goal. The most efficient way to do this is to let the individual teams concentrate and focus on solving one problem at a time where time is critical. If an organizational leader attempts to exert too much control over a battle or exercise, the soldiers on the ground will not be able to process all the information coming in, and time will be lost. Delegating responsibility and authority down to the team level will ensure that the teams move through Boyd's Loop quicker and more efficiently.

## TECHNIQUES OF LEADERSHIP

As with a new TL, the organizational leader should come in soft and quiet, unless it is obvious that the organization is in critical need of repair and that combat operations are imminent. Even then you should maintain the impression of being quiet, professional, observant, and forceful to set the stage for positive future leadership. As with the loud talkers and braggers that we previously discussed, leaders will be marked the first time they let their "alligator mouths overload their hummingbird asses." Troops will lose faith and confidence, and it will be difficult to keep playing a positive role.

## PROBLEMS OF ORGANIZATIONAL LEADERSHIP

One huge challenge when it comes to organizational leadership is finding a leader who has the guts to make the hard decisions in peacetime and just call a spade a spade. We are saturated with spineless leaders who do not demonstrate leadership. A leader must be able to look a subordinate or superior square in the eye and tell him what the problem is and what needs to be done to fix it.

I have watched too many leaders in both Special Ops and the regular army fail in this respect. If a leader cannot look a subordinate in the eye and tell him where he is screwing up, he is doing that subordinate a disservice.

The quote at the beginning of this chapter deals with another problem that combat soldiers face, and if you have spent any time in the military or law enforcement, you have seen it time and time again. Too many changes too close to the game indicates poor planning, preparation, supervision, and leadership. Failing to plan, review your plan, and to take input is likely to result in the failure of an operation. If a leader changes the plan prior to contact, it's either he doesn't have faith in the plan or he doesn't have faith in his men.

Another problem that can arise is the inability to make decisions. If a leader cannot make a decision in peacetime, he will have a problem when the stress of combat is looking him straight in the eye. I have witnessed this in instances where violence of action is required in the battlefield. Leaders who hesitate or don't use all their firepower and assets in training will hesitate to use them in combat.

Also be wary of the one-hit wonders. The policy and philosophy that states "You did it right once, and so you have punched the proverbial ticket, and then you don't have to consistently perform again" applies to this group. As I have stated before, selection is a never-ending process, and this rule should apply equally to both the guys on the ground and to the leaders supervising them. If an individual at team level screws up, he can get his buddy killed, or worse, his team. If a leader screws up, he can get multiple teams killed. Why in the hell would we keep promoting a leader that continues to stumble, especially in the tactical arena? Sadly, this is often the case.

I have also found that a strong base can allow a weak leader to survive. A weak or incompetent leader can come in and run a great organization into the ground for two years and then leave for another position. Typically, the organization, due to its members' professionalism and work ethic, can inadvertently allow this leader to succeed and then be promoted to a position of greater influence, where he or she can screw over more people. This is why leaders should look down and take a keen interest in their subordinates and not look to hitch their wagon to the next star.

By the same token, leadership in the military is generally a dual role. I will use the company level for my example. The company leadership consists of a captain and a first sergeant that should work closely together. Success for the company can simply be ensured by having one strong leader in either position. You can have a strong captain and a weak first sergeant, and the company will survive. You can have a strong first sergeant and a weak commander, and the company will still pull through. If you have a weak commander and first sergeant, the company is doomed. This is probably not too much of a problem in peacetime with the routine admin and exercises, but combat is another story.

I watched one company with two weak leaders crumble during combat conditions. It made me sick. The leaders, without a blemish on their records, were then promoted to higher levels of incompetence. This amounts to a form of protectionism.

I have witnessed protectionism of individuals at every level of leadership, including the team level. I remember a sergeant major who kept one of his buddies along throughout his military career, and even into his civilian one. The individual in question was a good man, but he had the leadership skill and personality of a rock. This guy could piss off a nun with little effort and brought a black cloud to the morale of whatever element he took charge of. First it was a team, then it was a platoon-sized element; then as a civilian he would piss off everyone in a 360-degree bursting radius. It amazed me that this exceptional sergeant major carried this blight along, but then again, the sergeant major did not have to work for him.

The same runs true with the officer corps. West Pointers are a group that have a clique within a clique. No matter how stupid the individual is, it seems to me the army sees fit to shuffle them around the service and not out. This goes back to my basic rule about being able to make a hard decision in peacetime. If you can't make a hard decision, you're part of the problem. You should probably do a quick check between your legs to see if your manhood or womanhood is still intact. If it is not, you're probably too big of a wimp for the job and should get the hell out.

Religion is another problem that I have encountered over the years that has conflicted with successful leadership. I have had commanders that brought religion into areas of command where it was best kept out. We had a commander that overtly promoted religion and had "prayer" breakfasts in the service, generally one hour before the normal meal. One positive thing it did was to get the chow hall open an hour earlier that morning for us heathens. Drifting into the mess hall early, you got a chance to see the "in" crowd. The few that attended this service were the bootlickers and cheese eaters that were intent on sucking up to the commander and being part of the "in crowd." You had some that were honestly trying to find some spirituality and apply it to their lives. Then you had some that were over-the-edge religious zealots who attempted to push their beliefs on you and anyone in their bursting radius.

I will be short and brief on my views of God. I believe you were put on this earth to do the best you can with what He gave you. To take care of your family and your fellow man as best you can. I believe one should not use God as a crutch, as many do, or as a political tool or discrimination factor in military job appointments. Spirituality in a person is best manifested in actions rather than spoken words. I had the opportunity to see an overly religious leader crumble in combat, caught up in his religious beliefs, almost like a spiritual Boyd Loop. My advice to men going into harm's way is simple: Pray before the fight or after the fight, but during the fight, you fight.

Another leadership problem that I have watched manifest itself is the "general factory." As Special Operations grew, so did the number of leadership slots at the top. Officers became more interested in looking toward the stars, the ones to be placed on their shoulders rather than down toward the men who would help earn those stars for them. It was almost heart-breaking to see incredible leaders who had been fire-breathing performers in the past, squelched and beaten into submission by the need to conform for promotion. Many feel the need to make general, as a measure of success. I won't go as far as saying that they sold their souls, but it was damn close. I feel some of them gave up some of their principles, or "bricks," to get the stars. My only hope was that they were able to find a glimmer of self-respect and that they attempted to do good in their new position of power and influence instead of continue to backslide and compromise their integrity.

How long should we let organizational leaders stay at the tactical level? I believe three to four years is a good start to ensure the tactical leader is well seasoned and fully understands the responsibilities and has seen most of the tactical scenarios play out. Proper leadership takes energy, and if the organization is properly managed, three to four years should be the right amount of time to rotate a leader and ensure he gets a physical and mental break. Staying longer can result in burnout and regression.

## WHEN TO SAY NO

I believe that leaders who lack the ability to project honesty and candor at all levels will, in time, jeopardize the health and welfare of their men. With all their education and continuing education, leaders need to look at the mission and beyond. They need to look at the *totality* and results of the mission. Somalia is a prime example of a country and a mission that we should have said no to. Somalia was a backwater nation where the United States had no strategic, economic, or political interests. Had the TF Ranger missions been successful, would the outcome have been any different? I think not. I personally believe that some countries need to evolve on their own and that we cannot fight their internal civil wars for them. Military leaders need to be smart enough to realize this.

A simple question that political and military leaders should ask when sending troops into conflict: Is the cause worthy enough to send their own son or daughter there to fight and die? If it is not, then they should weigh the cost of sending American forces. If the booger eaters in that region are at a point of social development where all they know is killing, nothing will change with your presence. I have found that they will stop fighting each other and start fighting you. Eventually, you will have to kill most of the problem individuals off to effect any change. Is it worth it? Will the American people support it? I don't know. As far as I'm concerned, the target country's people are an indicator: If they will not fight for themselves because they do not have the guts, the internal drive, and belief in their cause, why should we do it for them?

What I am saying in so many words is that our military leaders should not blindly follow orders; rather, they should look at the mission and beyond, and ensure that they voice their concerns loudly to our political leaders. If the political leaders fail to listen, the military leader should be prepared to use overwhelming force to protect their troops once they are put in harm's way. More times than not, commanders weigh the political atmosphere versus the hostile atmosphere when deciding how much force or support to give their soldiers going into the arena.

## TEST, EVALUATION, AND INTEGRATION OF NEW EQUIPMENT

With the flood of new equipment and the ability to procure them, leaders are now forced to pick and choose "off-the-shelf" items that will ensure combat success and survival in the battlefield. Want versus need must always be weighed, and solid tactics must not be replaced by gadgets. If it helps you do your job, fine. If it is just to make you look cool, dump it. As we used to joke, how high is the CDI (chicks dig it) factor?

During my time in Special Operations, many believed that we had an endless well, and whenever the spigot was turned on, new gear would appear. Nothing could be further from the truth. I wore the same heavy Kevlar vest for ten years. Fifty percent of my tactical gear was modified to fit the mission. A key item that we requested for years, and which we didn't receive until just immediately before combat operations, were NVGs. We had asked for these after our last major national conflict, and commanders looked in other places to spend the money. Finally, shortly before deploying on a live tactical operation, we were issued new goggles. The problem was that we were unable to train with them prior to deployment, and we were unsure of how they would hold up in combat. So we found ourselves modifying helmets and fabricating brackets for our helmets while deployed in a foreign threat country. This is not the smartest way to do business. All the while, the army was spouting the phrase, "We own the night." Maybe we just rented it for a while.

Well, we finally got the systems up and running, and they worked well. Because we had a finite number of these goggles and not any to spare, we elected to not take them on daytime missions because we were afraid of crushing or damaging them on a hit. Eventually, we modified pouches for them with pieces of ethafoam for padding. In later years, we were provided with nice molded plastic carriers that kept them intact during the roughest of training and combat.

## ENSURING THE COMMUNICATION PROCESS IS WORKING BOTH WAYS

Effective leaders do their best to ensure that a clear, quick, and efficient communication process is always available. The chain of command is the routine system used to deliver information to the masses. This ensures that each member in the chain is kept informed of changes and information while it is being passed down to the lowest member of the organization. Some information is so important that it must be passed en masse to ensure its timeliness and accuracy. Some leaders choose to put the information out en masse to ensure it is accurate and that it is interpreted in the correct manner with few changes or personal spins on it as it passes down the chain.

This should also apply from the bottom going to the top. I have witnessed some great tactical solutions which came from the bottom rung of the ladder. My job as a leader was to promote and exploit all the brainpower of my subordinates. No matter where the idea comes from, if it works, use it.

## KEY POINTS

- Keep your leadership selection neutral.
- Learn to counsel subordinates properly and honestly.
- Take decisive action when a subordinate's action (or inaction) requires a response.

# 6

# COMBAT LEADERSHIP

- **COMBAT LEADERSHIP PERCEPTIONS AND REALITIES**
- **KEEP IT SIMPLE**
- **MIND-SET: COMMIT TO THE SLAUGHTER**
- **TRAINING VERSUS COMBAT LEADERSHIP**
- **BOYD'S LOOP AND EFFECTIVE COMBAT LEADERSHIP**

## SCENARIO: TAKEDOWN

We received an intel that the number two man in the organization was spotted being driven in a lone green Fiat sedan. We quickly notified and briefed the force and launched the package. Our package had practiced countless hours on takedowns of this type, and we were ready to interdict this or any other vehicle configuration.

Per the plan, the lead bird with snipers spotted the vehicle and engaged it. After a magazine and a half of semiautomatic fire into the engine compartment, the vehicle finally expired on a side street. Its occupants fled in a panic to a nearby house. The homes in this area were upscale, with heavy metal gates and thick masonry walls surrounding their courtyards.

The birds deposited us about one block away, and we proceeded to look for the target vehicle. After the initial dust cloud from infill drifted off and the birds departed, we began our movement down the street. Approaching the target intersection, I saw a man lying on the ground next to the wall. The lead teams had bypassed him. He looked to me like a beggar, wearing somewhat-raggedy clothes, possibly a cripple that you would see from time to time on the street corner. We looked left and saw the empty Fiat lying on its side in the middle of the street, bullet holes in the hood.

We ducked into the first courtyard on the left and observed teams clearing the residence. Having enough teams to clear the target, I told my guys to drag raggedy man into the courtyard for safety and to flex-tie his hands behind his back and then tie one of his feet to his hands so he would not run away or interfere with our assaults. They promptly did this, and when they grabbed his left foot to tie it to his hands, his whole leg flopped, and his foot damn near hit him in the back of his head. My limited knowledge of the human anatomy and the little voice in my head said that was probably not good. I called for a medic to check the man out. It turned out that raggedy man was the driver for the bad guy we were trying to capture. Evidently, he took a round in the upper femur, which severed the bone. He did not make a sound during this entire episode.

We had lost eyes on where the vehicle's occupants had gone, so we decided to start clearing each building in the surrounding area. We promptly turned raggedy man over to one of the dedicated medics and headed for the courtyard of the house closest to where the Fiat had crashed. We pushed in through the heavy metal gate, and I told Tony to take the guys and clear the two-story residence.

I pulled security at the inside of the front gate, and I wanted to keep tabs on where all the teams were, including the support elements. I also wanted to ensure no one back-doored us on this target. It appeared that everyone had bomb-burst in all directions away from the green Fiat in an effort to pin our target down and not let him get out of the area. From my position, I could see the command element being deposited on the roof of the first residence, where raggedy man was being treated in the courtyard.

The commanding general, watching this action unfold in the rear, reportedly commented, “How do they know where to go?” It is actually easier than it looks. The bad guys are on foot, and they can only move so far so fast. It is a matter of eliminating all possible hiding places as quickly as possible. The ability to take the initiative at team level is what ensured our success on this operation.

AK-47 fire was starting to pick up, and an RPG impacted on the outside wall or building next to where I was covering. It rocked the immediate area and rang my bell. Looking down the street away from the Fiat, I could see a woman in a flowery orange dress standing on the left side about a block and a half away. It looked as though she had a gunman shooting from between her legs, as I could see the dust kicking up from the muzzle as he shot. I thought about tagging her, but I did not know where our support forces were, and I figured if she was a threat, they would take care of her. Little did I know our blocking force was still circling overhead and had not been alerted to this hit. Why? I hadn’t a clue. I learned long ago in training that if you’re not sure of your target or where friendly forces are, you should not shoot.

Tony brought the team back to me, and he had recovered an AK. The building had been empty of personnel. We quickly stripped the weapon and threw parts in all directions. Scanning the street, I could see that the command group was pointing to a building across the street from me. I could see Ben occasionally pop up and shoot at something in the distance and then point to what appeared to be the building across the street.

I was trying to figure out what they were pointing at and was hoping they would throw smoke on the target as we had asked them to do in the past. No such luck. With sporadic AK rounds still flying, including an occasional RPG, we darted across the street in pairs without laying down cover fire. We thought our security forces were out and did not want to accidentally shoot into one of their blocking positions. So we held our fire and sucked it up and made successful runs in buddy teams.

As I got to the double courtyard gate I saw that it was hastily tied shut with a piece of rag, and this made me think we were getting close to our target. Another team had joined us on our left, and as we pushed through, they headed for the front door. Seeing this, we headed for the back to ensure no one snuck out.

The building was a nice two-story structure, and as we rounded the back corner, Kim advised that we had company. They were looking out the back door, or what appeared to be a hole in the wall of the residence. I told him to bang it, and we threw a diversionary device in. It went off in a second and a half, and then we entered. The banger had done its job; we saw eight people in a small room. The concrete walls and the roof of the structure had only amplified the concussion of the flashbang, and everyone appeared to be dazed. The team started putting everyone down, and I pulled rear security at the door for a minute because I would just be in the way with all the bodies and the clutter. While the guys were going hands-on, it was my job as TL to ensure that we had overall security and that the team was focused on the task at hand.

Our sister team pushed through the front of the target building and linked up with us. Other teams had converged on the target and were pouring in and heading upstairs. We quickly secured the building and began to consolidate everyone in the front room for identification. The front room was a larger room with better light, and we used it as a common meeting place to link up after the target was secured. As we started to consolidate our prisoners in the first room, a member of the consolidation team started questioning one of the guests. The person in question blurted out his name to the operator. Bingo, we had him.

We notified the command element that we had our man and requested an exfil. We were advised to use our roof as an exfil point, and the birds began to shuttle in to retrieve our teams and prisoners. We were still taking fire, and in the lulls we looked hard through concrete latticeworks in the stairwell for someone to shoot at. We could hear the gun birds working some folks over with rockets and miniguns and knew things were heating up. As the birds came in for us, we positioned with our teams and quickly hopped onto the bird with our prisoner to cut down the exposure time.

We lifted off and did our best to look for targets to suppress. We could not see any, but we could hear the AKs hammering away at us and our positions. The bad guys were tucked in tight wherever they were. I heard that some were even in trees, but I could not spot any muzzle flashes. It sucked because you wanted to shoot, but you knew you had to maintain fire discipline. To my knowledge, no one on our bird fired because they could not identify any targets and because we did not know where the blocking positions were.

Once we were out of the target area, things quieted down, and our pilot deposited us at the front gate of our compound. We secured our prisoner and handed him over to a prisoner-of-war team, who would search him, photograph him, and prepare to interrogate him.

I got my team together and began our personnel and weapons check. This is routine after an action to ensure no one is hurt or has been shot but does not know it. With all the adrenaline, guys would not know they had taken hits until after the fight. Besides checking to ensure you had all your equipment, it's mandatory to check yourself and your buddy out.

After this quick check, I would get the team's story straight. This is not what it sounds like. When Tony took his target house down, I had no idea what they had done, found, or if they had killed any gunman. This was my time to find out what happened out of my sight so I could give an accurate report to the commander. I did not want something to have happened and me be not aware of it, good or bad.

We entered the debriefing area, pumped up and happy to be alive. A couple of guys high-fived, because they were happy to see their buddies were safe and alive. Between snipers and gun birds, they had engaged over twenty bad guys during the hit, and this had been the most fire they had taken to date.

Our commander chastised us, stating, "You men need to settle down. There are some families out there without their loved ones." He was referring to the bad guys who were trying to kill us. Emotionless, we looked at each other, and you could see the truth in everyone's eyes. We all felt that our commander was near his psychological and emotional edge and that his religious beliefs were getting in the way of his combat decisions. We had feared this would come to pass because of his history. Our commander was a good man, but missed his calling as a chaplain. What was worse was that it appeared that his deep religious beliefs were interfering with his combat decision-making abilities. This would play out later on a large scale during more critical operations.

## AFTER-ACTION COMMENTS

### *SUSTAIN*

- Team leader aggressiveness and initiative.
- Discrimination and weapon employment.

### *IMPROVE*

- Better communication as to who is on the ground.

## COMBAT LEADERSHIP PERCEPTIONS AND REALITIES

Many leaders, both NCOs and officers, assume the leadership role with many preconceived notions of what a leader is and how they are supposed to act. We have all seen the typical movie roles of the leader out front, leading the bayonet charge into the enemy. It is a technique, one that I don't really recommend. Effective combat leadership should empower the fighting troops to take initiative and decisive action at the lowest level to support the common goal. Leaders that have properly trained troops and have confidence in their tactics and capabilities are not afraid to do this. Not all troops are up to this level, but this is where we should strive to be.

One must understand that in today's battlefield, we are routinely fighting a "G" (guerrilla fighter) who is lightly dressed in what we would term pajamas, has only an AK-47 and two magazines for it, and can run circles around us in our heavy combat gear. We must focus on getting to the better position first before he does, or he will have the advantage. The shooters at ground level must have the latitude to close in on and engage these guys on our terms and not theirs. Should we see a weakness, we need to efficiently push through them to superior positions and then call back to our next level of leadership and brief them on our progress. If not, the G will get into the better position first, and we will have to try and root him out of it, which puts us at the disadvantage. Also, one must understand the window of opportunity to engage a target in a fast-moving gunfight, which starts and ends very rapidly. You are either mentally, tactically, and physically ready to take the shot, or you are not. This applies to the lowest level of the team.

## KEEP IT SIMPLE

Don't overcomplicate the command and control or checks and balances you use on the battlefield. Simplicity is the key. Remember to rotate your subunit personnel and missions to give everyone a chance at planning and mission leads. This will keep everyone sharp and focused. Also, keep the lines of communication open and have backup plans should the primary means fail.

It is critically important that your soldiers know the capabilities of their weapon systems, their safe use and discrimination. With the upgraded lethality of our weapons systems, we cannot afford to have a bad day. I can remember being on several operations and taking fire from unknown personnel. On several of those instances, I could not identify who was shooting at me. By engaging the threat, I would put other friendly troops in jeopardy, so I did not fire. Sometimes you need to find hard cover and look hard during fast-moving gunfights to ensure you're putting rounds in the right people. This goes for Special Operations forces as well as for conventional forces in today's chaotic mechanized engagements.

## MIND-SET: COMMIT TO THE SLAUGHTER

As discussed earlier, leaders must be as equally stable as the soldiers they are sending into combat, and they must have the same aggressive mind-set, but on a larger scale. They need to be mentally ready to commit slaughter on several different targets or in several different areas at the same time. The individual or the TL is going to neutralize threats as they come, usually one at a time. The leadership, on the other hand, is going to be watching this from a command position, a bird, or even on a television screen.

The leader's mind needs to be right and focused. He may be tasked to target suspected enemy positions or civilian positions that the enemy has taken to engage his men. Whether or not civilians are there, those positions need to go away. This seems a simple matter to me, but leaders too far from the fight will put too much thought into this simple scenario. It gets even more complicated when we have to preemptively take out targets that we know are bad. We will have civilian casualties. Such is the case where the bad guys put an air defense site on top of a civilian apartment building. I have a simple rule for this:

American Servicemen First

Don't worry about the press, political considerations, or what the neighbors are going to say or think. Take a leadership position and accept that you are responsible for your soldiers' lives and accept the fact that you must be willing to commit to the slaughter of anyone who threatens the lives of your team or force. It is our duty to bring all of our personnel back home alive if at all possible.

Currently we are infested with a generation of leaders who are not willing to commit to the slaughter. When American bodies were beaten, drugged, abused, and shown on television after the battle in Somalia in 1993, there was no response. I remember when four American bodies were burned, abused, hacked, and torn apart and hung as trophies in Iraq. The American military and leadership response was zero. Great rhetoric spewed from the mouths of our military leadership, but nothing happened.

We have the technology now, as we did in 1993, to send a message to these savages that their behavior is not acceptable. But first, you must have the balls to do it. Our leadership in 1993 did not have the guts to do it, and it seems that many of our current military leaders have the same problem. We could easily correct this behavior by sending a message that would echo across the world. What really sent this point home to me is that I have fought beside these warriors, and I come back to look their family members in the eye when they could not.

I would rather kill a thousand booger eaters than have to look another American wife, son, or daughter in the eye and tell them how their family member did not make it. The current crop of military leaders are conditioned to think of these as "acceptable loses."

I have never bought into the bullshit theory of acceptable losses. We have the technology, the talent, the aggressiveness, and the expertise to bring more of our warriors home than ever before. Our combat, military, and political leaders choose to be mediocre at their jobs. That is as simple as I can put it.

A sample of the thought process can be gleamed through the following example as e-mailed to me by a former cadet and now army officer. Another newly commissioned second lieutenant reports a similar account from Kosovo:

> *As you may or may not know, my platoon ended up going up to Mitrovica (Where the French are) attached to the Parachute Infantry Regiment from Bragg. The place is very hot as far as activity. Conditions were very austere as we lived at an old tire factory.*
>
> *My platoon's first mission was to conduct a cordon while the infantry searched buildings (at a college) for weapons in the Serb part of town. At first things were going ok as we started right when the curfew was over in the morning. A couple of Serbs even said that they were glad that we were here. I was also interviewed by Rugger's International News Agency and Radio Free Europe. But a couple of hours later,*

*the Search was over and they still had us cordoning off the area because the French were not done. Then some of the Serbs started to get angry because some of us had Albanian interpreters. One interpreter even had to be evacuated from another TCP (Traffic Control Point) that the Anti-Tank Platoon had because they were threatening to kill him. American troops were not kicking down doors as the Serbs claimed in the news. The United Nations Mission in Kosovo Police apparently had other plans and were kicking down doors. American troops did have people cut locks on doors which made the Serbs angry. Several weapons and a grenade were found. Then all hell broke loose everywhere.*

*Pretty soon we had a mob of over 3,000 Serbs that were hurling insults, rocks and bricks at us. We mounted our vehicles and I was waiting for the word to get out of there, because shooting someone would still not have been a good move at that point. I was spit on, my gunner was hit in the K-pot with a rock, another one of the gunners in my platoon broke a finger with a rock, another gunner's tooth was chipped, and the Serbs broke windows and mirrors in our up-armored HMMWVs.*

*We were completely surrounded and I finally told my driver and the squad I was with to go and have the rest of the platoon meet us by the east bridge which crossed back over to the South Albanian side. The Infantry was also attacked, but no one ever fired any weapons at us and we did not see any.*

*The French pretty much let all of this happen, but one French soldier actually pushed Serbs back away from me while I mounted my vehicle. The entire Search element, the AT Platoons, the snipers, and my platoon got back to the tire factory wondering, "What the hell just happened to us???" We literally felt like we had gone into a football game and got our butts kicked. LTC Smith, the BC in charge of us said that we had done our jobs, and we all maintained excellent discipline by not upgrading our level of force to something which would have been an international incident. I felt bad because I had told my gunner to stay up on the gun ready to shoot if the threat possessed itself, right before he got hit in the K-pot.*

*The following day at dusk, we marched across into the North side of Mitrovica to a neighborhood called Little Bosnia, because of the Serb/Albanian/Turk mix population. These people were much poorer, but the neighborhood was adjacent to the one that we had been attacked in the day prior. My platoon provided a screen for the search element. The whole operation took place without incident, but we had to stop the search, because we had gotten word that up to 40,000 Ethnic Albanians were on their way to Mitrovica from the South.*

*We went back to the tire factory and stood by as we waited for word to move. Over 40,000 Albanians from as far as Pristina had showed up with the intent of returning ethnic Albanians to the North side. They went up to the east bridge which was occupied by a Bradley platoon, some Brits and some French troops which ended up having to use tear gas on the Serbs that had gathered on the North end of the bridge because they felt that the Albanians were coming with weapons and KFOR was not doing anything about it. The Albanian Demonstration was peaceful, however. No one ever came to the East Bridge, which my platoon was occupying with the Germans and the French but the French troops were very nervous and closed off the bridge anyway (Albanians don't like the French because they believe them to be Serb allies). We held the bridge for over 7 hours and again we left with no incident.*

*The following day my platoon had to provide a presence patrol of an area that was completely bombed out. This area had no inhabitants and looked like what the future looks like in the movie* The Terminator. *All rubble, buildings barely standing, and streets empty. It was a significant area, because snipers had been reported just across the river on the Serb side.*

*On the last day, my platoon again provided a screen for the search element, this time in a completely Serb area and this time with less than lethal munitions and riot control gear. This time the plan was once the search was over, my platoon and another MP platoon that had been brought up the day prior, would lead the infantry through any Serb mob and make our way back across the west bridge. This time if they started throwing rocks, we would snatch them up to make an example of them. If they continued to throw rocks, they would be hit my 40mm M203 rubber bullets, foam baton rounds, and multiple foam baton rounds. If that didn't work or if they started shooting because they say us shooting, we would go to the real stuff… 5.56mm ball and tracer.*

*We really thought that we were going to have to shoot our way out. One concern was that if we started shooting the less than lethal munitions at them, their snipers would see us shooting and Serbs falling, not knowing that we were not shooting real bullets. Their snipers might then start shooting at us and we would have to start shooting real bullets.*

*When the search was finished, not enough of a crowd had gathered and the French were able to hold off the Serbs. We moved back across the bridge standing a little bit more proud, because this time we were leaving on our own terms. The Serbs were not able to get things coordinated that day. All they could do was hurl insults at us while we marched by. I am glad that no one had to die or get hurt that day, but I am also disappointed that I did not get to see how a 40mm*

*M203 rubber bullet could knock someone on their ass and make their world hurt.*

*My platoon accomplished the missions it was given, and the soldiers maintained excellent discipline and restraint, preventing any unnecessary loss of life. Keep in mind that not all Serbs feel that way towards Americans. The Serbs in the American Sector actually count on us for their protection. The Serbs in Mitrovica still don't like that we bombed Serbia for 78 days. My platoon may return, and if we do, we are ready to deal with whatever missions we may be given.*

*—Bubba*

In the narrative, I put the details of the commander's comments not as a personal attack but as a learning point. I have witnessed this type of behavior time and time again throughout my career, and it always causes problems. Individuals selected for Special Operations generally have a focused and determined mind-set upon being selected. Without it, they generally would not have made the selection process. Many find religion after their induction, and I do not have a problem with that as long as it is kept in its place. I have commented during my combat mind-set class that the time to pray is before the battle or after the battle, but during the battle you must focus on killing—the mission is killing the bad guys.

This is where the problem comes in. Military officers should be surgical to a point, but when it comes to the safety and welfare of their troops, they need to draw the line. Should you need more air strikes or artillery to effectively protect your troops, you should have your mind committed to this prior to the battle. Again, I would urge every soldier to ask their commander a question prior to combat operations: "How many people are you willing to kill to bring us back?" The answers will probably surprise you. As a soldier going into harm's way, you need to know where everyone stands on this issue. If leaders start back-pedaling on a simple question, they will most certainly do it when times get tough. As a leader, you need to do some serious soul searching and personally establish what price you are willing to pay to bring your men home.

## TRAINING VERSUS COMBAT LEADERHSIP

I've made this point before, and I will make it again: Whether it is an officer or an NCO, if a person can't make the hard decisions in peacetime, he will find it harder to do so under the stress of combat. Routinely force these leaders to make hard calls during training scenarios so as to visualize and work the solutions out prior to encountering a life-and-death situation.

Again, in Special Operations, we go through thousands of hours of live-fire, close-quarter combat training, entertaining every possible scenario and adjusting our tactics to simple drills that will apply to them all. Why? Referring back to the muscle memory equation, it takes two to three thousand repetitions to get it down. How about "mental muscle" memory? What difference do you think is there? Probably not much at all. So if a shooter on the ground is doing repetition after repetition of his common skills, should a commander be doing the same type of mental rehearsals? You bet. Five or six times in a command bird is inadequate.

Leaders need to mentally rehearse the various options that can play out and make the appropriate decisions. Yes, it would require a bit of work, but no more than we currently do with shoot / don't shoot scenarios on the interactive video training machines. Common combat scenarios could be played out and a common list of options drafted that would give the commander multiple solutions to a problem. Solutions that could be put away for future use in similar scenarios. Combat can be described in varying degrees or intensity. I have participated in training scenarios that were more intense and difficult tactically than many of the real-life situations I encountered. Combat takes many forms, but we should limit ourselves to one simple mind-set.

## BOYD'S LOOP AND EFFECTIVE COMBAT LEADERSHIP

The reason we should seek out simple techniques is that if we don't use them, we are setting ourselves up for failure on the battlefield. As technology improves, so must our ability to amass battlefield intelligence. I mentioned earlier that at the individual and team tactical levels, I used Boyd's Loop to describe how I process information. The same must be done at the command levels. With so much information being input, commanders need to filter out what is fluff and focus only on what is critically important in reference to their current tactical situation.

## KEY POINTS

- Ask your leadership a simple question before conducting combat operations: "How many people are you willing to kill to get us out?"
- Ensure that religious issues are addressed prior to combat operations (see above question).
- Remove leaders who cannot make the hard peacetime decisions. They will fail in combat.

# 7

# TRAINING FOR THE FIGHT

*Above all, we must realize that no arsenal or no weapon in the arsenals of the world is so formidable as the will and moral courage of free men and women. It is a weapon our adversaries in today's world do not have. It is a weapon that we as Americans do have. Let that be understood by those who practice terrorism and prey upon their neighbors.*

—Ronald Reagan

- **LEVELS OF PERFORMANCE AND TRAINING**
- **THE PATH OF SUCCESS**
- **TRAINING RESPONSIBILITIES AND SEQUENCE**
- **TRAINING PITFALLS**
- **REHEARSALS**
- **AFTER-ACTION REPORTS/REVIEW (AARs)**
- **MAINTENANCE TRAINING**
- **EVALUATIONS**

## SCENARIO: THE RADIO STATION

It was a near miss. We tried again to pinpoint the target, but with no success. The intel guys would continue to try to provide us with timely and reliable information, and we would continue to rehearse. We had not worked with a vehicle assault configuration in years, so we developed a simple plan using some of our security personnel and Humvees. We were assigned a driver, a track commander (TC), and a gunner. The driver's job was simple enough—put the vehicle where we needed it. The gunner's mission was just as simple—shoot at any bad guys that presented a threat to our force. The TC's job was a bit more difficult. He needed to control the driver and the gunner and know exactly where the vehicle was at all times, logging in checkpoints as we drove. Should we make contact, he and his crew would have to act in concert with the other vehicles to ensure that swift and decisive action was taken to get the "hell out of Dodge."

Our vehicle's TC was a good man, and we worked well together. We planned out how we were going to load the vehicle and what actions we were going to take on contact. We established fields of fire for every person in the vehicle and made it understood that one side might be in contact with the enemy and the other side must still watch their sector and ensure that the bad guys were not going to hit us from both sides at the same time. This took discipline and focus, as your partner with his back to you might be actively sending rounds downrange, so you could not afford to turn around and drop security of your sector. You had to trust he was hitting his target and that the lead and trail vehicles were helping him out with fire from their vehicles.

While waiting for dark to come, we prepared and double-checked all our personal gear to include gun lights, markers, personal lights, emergency strobe lights, and lasers for our weapons. Once everything had been checked and spare batteries were put in the gears we checked the vehicles out. Sandbags were placed on the floorboards and beds of the vehicles to stop or slow down shrapnel from mines or improvised explosive devices (IEDs) placed in the road. We had seen videos of vehicles cut in two because of these homemade improvised mines, and they were just as deadly as military-made ones. We made sure we brought the team breakaway bag, a small rucksack packed with extra ammo, bandages, and water. This was stuff you might need if you had to leave a vehicle should it become disabled. This would also give you a bit more ammo to sustain yourself should you need to stand and fight for a while.

Once the vehicle had been checked and double-checked, we quietly assembled around it and reviewed any last points prior to our movement time. When the time came for our rehearsal move, the TC and I walked around the vehicle to ensure that all my guys were up, aboard and ready. I then took my spot behind the TC, who was seated in the right front passenger seat. This way I could talk to him almost face-to-face during the movement. The vehicle's TC gave the ready count to the convoy commander, and we began our movement around the camp. We probably drove ten kilometers that night, allowing the drivers to get used to following other vehicles, keeping pace and maintaining the proper distance. We moved the vehicles into a staging area, where we dismounted and conducted a dry run on a simulated target. Once this was completed we got accountability of our teams and forces and then waited by the vehicles in security positions before giving the word to remount.

Once the go-ahead was given, I, as the TL, was the last to remount. I needed to see that each of my guys was on the vehicle. I would then pass the word to the vehicle TC, who would again relay the information to the convoy commander. We would then begin our movement. It was critical to ensure that this remounting process was swift and that we could begin moving again. Speed and movement was our security, and none of us liked to be cooped up in a vehicle that you could not effectively maneuver or fight from. One knucklehead with an AK could put a great deal of effective fire on seven guys stuffed into one HMMWV (High Mobility Multipurpose Wheeled Vehicle). Even a bad guy that was not that accomplished a marksman, some rounds were bound to connect with someone.

We returned to our staging area and quickly debriefed on the actions as a team, and then I went on to the main debriefing. The plan was coming together, and the movement was becoming a bit more polished. The word was passed around that if the hit did not go tonight, it would probably go tomorrow night. We decided to do another movement and rehearsal to give all of us a chance to work out any bugs and get our plan down pat.

The major difference in this plan was that we had switched the teams responsible for assaulting and blocking positions to give the blocking teams a chance to work on their assaulting skills. Originally, my team's mission was to assault the target building. Now it was to provide a security position for the newly assigned assault force. I thought it would be a much easier job, but things were to get busy on this hit, which was to become a bit more challenging.

As it came to pass, the location of the target was found and pinpointed. Intel and vehicle commanders planned a route to and from the target area. They started from our assembly area and passed through a friendly foreign national–armed checkpoint, a place that is routinely dangerous.

Passage of friendly lines was always a very dangerous and demanding module taught in ranger school. Anything could happen, as it was the final line you crossed before you went into Indian country. Artillery could rain down on you prior to the passage, or the enemy could simply attack that area or sector. This was not that big of an issue. But if the enemy attacked while you were crossing the final friendly point, things could get dicey. You had the choice whether to continue your mission or to return to friendly lines. The problem is that everyone could be shooting at this time. Friendly forces on the line, bad guys, your people; it could become a mess very quickly, especially at night. Compound this with not speaking the language of the friendly forces who own the checkpoint, and you could have a major headache. We left one of our unit members as a liaison with these friendly forces so they would be there on our return and could tell the guys behind the guns who we were.

The mission was planned to be carried out slightly past midnight. We had a routine day ensuring that everything on our vehicles and our person were checked and double-checked. As the sun set, we tried to get some sleep, but that is almost impossible. Your mind continues to race, and as you lie there and try to rest, you review the plan in your head to ensure you have taken all the steps to ensure your success. At a predetermined time, we woke everyone so they would have a chance to get some food down, take the final latrine call, and generally just wake up and get focused.

The team was ready, and we moved toward our vehicle. I linked up with the other TLs and met with the AC to see if any other intel had come to light. There was none, and we returned to our vehicles, mounted up, and waited for departure time. As we waited, the occasional team member might dismount and take time for that last piss. The last thing you want to think about when going into combat is the pain of a full bladder when you're looking for bad guys. Once we were moving, we were not stopping for shit. If you got hit, we would not stop to treat you unless we were back home safe.

The time came, and the signal was given to move out. The lumbering convoy picked up its cruising speed and headed toward the passage of lines. This was uneventful, and we did not even stop. Our liaison was there, and we rolled on through, lights blacked out, and ready to do business. As we moved into the city, we noticed that it looked like a concrete ghost town. The lead vehicle was reporting a few individuals on the street, most of whom were unaware of our presence until the lead vehicle was right on top of them. They would quickly duck behind a post or a tree and wait until we had passed.

The lead vehicle was equipped with suppressed guns and had the ability to neutralize any threats they encountered. All the people on the way to the target did not want to play. Evidently, this was the first time a group or unit had ever done any night operations, and the locals were probably shell-shocked. Most did not know of our presence until we were face-to-face and they had no time to react. The diesel engines can be extremely quiet if you learned not to rev them, slowly bringing them instead to an intermediate cruising speed. Our flow and movement pace was going all too well when we hit a snag.

The lead vehicles ran into a string of concertina across the road. This stopped the entire formation. If driven over, concertina wire will wrap around your tires and axles and eventually shred your tires, all the while dragging the remainder of the roll behind you. Nasty stuff.

The other problem of encountering an obstacle such as this is the possibility that it is covered by fire. This is the perfect way to stop an entire convoy and allow multiple gunmen to wreak havoc on a thin-skinned vehicle formation. In addition to gunfire, you could easily put an IED all up and down where the lead vehicle was and take out several vehicles with one push of a button. Fortunately, it appeared that someone had pulled the razor wire across the road, left it, and gone to bed.

As our movement continued, code words were given, letting us know that we were getting into the target area. As we made our final turn onto the street that contained our target building, we started looking hard for the first intersection on the left. We were to turn down it and travel to the next intersection and set up a blocking position to protect the assault force. Two vehicles in front of us passed our intersection and headed for the next street.

I found the intersection that led to our position being blocked by a vehicle. Not some little Fiat or sedan, but a fricking garbage truck. I just shook my head and told the team to dismount. We were going to have to do this the old-fashioned way, on foot. I hated to leave our vehicle behind and the heavy weapon mounted on it, but I had no choice. The assault was going to happen, and we had to get to that intersection to keep people safe.

The team moved in pairs, and we drifted past one major opening to the courtyard on our left and passed several closed doors on the right. As we approached our intersection, we set up in the open in buddy teams, each taking a street to their right or left and a street ahead of them. It was quiet for a few moments, and then the silence was broken. One of the Black Hawks containing snipers buzzed low, fast and loud about fifty feet over the target, and woke everyone in the neighborhood up. *Honey, we're home.*

I was fucking pissed.

It was an exceptionally clean and surreptitious insertion up to that point, and then everyone and their brother woke up. The door opened on the corner building we had our backs up against and someone poked their head out. Scott and Tony grabbed a local, threw him down, and flex-tied him. All of a sudden, I could see heads poking out up and down doorways on both sides of the street. In addition, the assault team had thrown their first distraction device, and the sound echoed throughout the neighborhood.

I could see the assault force enter one of the doorways we bypassed, moving toward our position. A moment or two after the assault teams entered their breach point, a man appeared from the opening across the street from their entry point, turned, and started walking my way. As he did, he half-concealed an AK-47 at his side, the distinctive half-moon magazine giving it away. Not having time to report him, I started to sparkle him with my laser. I knew if I were to engage him now, the bullets would pass through him and go into the blocking position less than twenty-five meters to his rear. So I waited. He did not see us and kept walking right toward our position. He continued to walk, and then another man came out of the same opening, following the first man's route.

As the first man got about twenty feet away, he must have caught sight of our silhouettes. He kept walking straight ahead and tried to pass the AK to his other side away from us. My thought was that he was going to try to make it to the corner and either get away or try to engage us. I could not take the chance of the latter happening, so I held the laser on him center mass and waited until he was perpendicular to me about twenty feet or so away. This would ensure the rounds going through him would impact the wall directly behind him. Correcting for the offset of my laser, I began to engage him with some 5.56 tracer. I watched as the tracer rounds went through him, hit the wall, and bounced off.

In my mind, I stopped with a double tap, firing the two shots that I had always been trained to do. The problem was that he was still standing. This probably lasted only half a second in my mind before I continued firing with two more rounds, at which point he fell backward into the stream of bullets. I started to close in on him, and to my amazement, he stood back up for about a second and then collapsed back down on his back.

I began to scan back down the street for the number two man who had been following him but he was now assholes-and-elbows running back through the hole where he came from. I quickly grabbed a flashbang and threw it hard over the courtyard wall in his direction to let him know that I knew he was there.

Turning my attention to the man down, I told Kim to grab the AK and to cover me while I checked on the man. I knew in my mind that he was dead because my weapon was zeroed on him, and I could never miss at that range. The man was attempting to breathe, but he did not appear to be getting any oxygen. I soon found out why. Searching him, I found that his jacket on his left side where the bullets had exited was heavy and full of blood. He was fighting for air, but could not get any. I almost felt sorry for him as I cupped his head with my left hand while I searched him with my right. His eyes rolled back, and he expired on me.

Up to this point, I had blown two team SOPs in dealing with this bad guy. As a TL, I should have made Kim get hands-on while I covered, and I should have flex-tied the bad guy prior to searching. I got lucky. During my search, I found one of those key chain noise makers that sounds like a Star Strek Fazer. The label on it said Echo-1. This was my team designation, "E" team, and this was my designation on the team, E-1. I thought this was fucking spooky, and I left it on his body. I thought for a moment about dragging him down the street and throwing his body on the hood of our vehicle, similar to that of a deer, but I figured that it would not look good if someone got shot dragging him back and that the driver would probably have a hard time seeing over him. We got the word to remount, so Kim brought the AK with him, and we moved back down the street to our vehicle.

Our order to remount our vehicles was a bit premature. The assault teams were still doing business on the inside. I grabbed Kim with his M203 grenade launcher, and we headed about ten yards back down the alley. I found a dirt pile on the left-hand side, and we took up a cover position there. I figured that Kim could put some high explosive rounds on the intersection where we left the bad guy. When Pete came out with his team, he looked to right and could see the body I had left on the street and no blocking team. I think he also realized that we had already been pulled back. Quickly, he brought his guys down our street and to his vehicle. Soon we had all loaded and were starting to move.

Tension was high as the bad guys in the area knew now where we were working. We were all watching our sectors when the vehicle behind us spotted some bad guys in an alley getting ready to light us up. A team member from another vehicle engaged them with a squad automatic weapon (SAW) and made them dance a bit. Scott was also able to put some fire on them before we moved out of sight and range. Our air cover started to work over an area with rockets and miniguns, which made us feel better. The problem now was getting back through friendly lines without getting hit by friendly or enemy fire during the process. To my relief and amazement, we breezed through without incident.

We dismounted our vehicles, and I did an inspection and debriefing and told Kim to turn the AK over to supply. I wanted to ensure that the weapon was properly processed so that I had some physical evidence as to my reason for firing. The command debrief went well, and we were able to catch a few hours of sleep.

### AFTER-ACTION COMMENTS

*SUSTAIN*

- Continue aggressive rehearsals.

*IMPROVE*

- Individual searching and flex-tying.
- Perimeter collapse procedures.

---

## LEVELS OF PERFORMANCE AND TRAINING

While serving in Special Operations, I was able to work with a host of instructors from various tactical units foreign and domestic, large and small. I quickly learned that levels of performance and training vary from organization to organization and unit to unit. Through the years, I had the opportunity to observe training and exercises at all levels ranging from small-town teams to foreign counterterrorist units.

Individual, team and organizational level performance and capabilities are directly tied to exceptional leadership. The common thread that ensured exceptional training was an exceptional cadre responsible for the training. How do you get there? You invest. By investing quality people in training programs and ensuring that they have the resources to pass the needed information down.

A problem area that needs to be addressed is one that I term "poisoning the well." This describes the trend where training sections, organizations, and academies are used as depositories for unwanted or substandard individuals. If I were a new police officer reporting to the academy, my initial view or perception of an organization would be greatly shaped by the training cadre. If I saw a group of unhappy, sloppy, fat staff that could not get along, were unprofessional, did not believe in the organization, or were unhappy with their job assignment, I would soon develop a skewed or tainted view of the organization myself. This would be especially true if I had to spend three to six months working closely with the same staff.

I was fortunate to have served with a crew of instructors that set the exact opposite standard. While I was in Special Ops, the command supported the selection process and assignment of proven instructors to our training section. It was an incredible experience. Potential instructors were selected from individuals who had successful TL time under their belts.

During my tour as an instructor, my fellow cadre members had combat experience. If you were a new instructor, you were paired up with another instructor, and you worked with him while he taught his assigned blocks of instruction. As an instructor rotated out, the new instructor became the primary instructor for that block of instruction and had the latitude to change or improve the material.

I noted that every instructor made changes without guidance. They made their presentations and training a little better for the new instructor who would take over the class. This created an exceptional work environment for a cadre seeking perfection in their duties. It also created a united cadre that supported each other, a situation that was readily observed by the students. There were no chinks in our armor, and the students knew it.

The students quickly received the message that they were there to train and not bullshit around. They had a cadre of proven professionals who would put the same gear on and do the same physical events as they would, setting a training standard and a system that I still employ today. No one complained or even thought about complaining. Some of these instructors were fifteen to twenty years the senior of their students. I remember when I was going through training and the cadre took us on a seven-mile run carrying an M14 and combat-loaded load carrying equipment (LCE). Our instructors were there right beside us the entire way, enjoying the North Carolina summer heat.

What level or goal should you expect your team to attain? This will differ from group to group. Training time, facilities, and equipment play an important role in this equation. Of course, safety should also be a primary concern. This primary concern should focus on not injuring a student through negligence or stupidity.

Later in the book, I will go on to describe my training tactics and techniques in more detail. Here I hope to outline the general principles that form the foundation for a successful training program. This general overview of training principles is something I developed over the years both as an active serviceman and as a private instructor. Before anyone can get to the nitty-gritty of training techniques, he or she needs to grasp the overall ideals of training as described below.

## THE PATH OF SUCCESS

Individual skills and tactics must go from dry-fire (practice), to range fire, to paintball scenarios or simunitions with live role players to combat on the street with little or no changes. This philosophy of linear progression will enable you to have a valid system of training that will ensure that you climb the tactical ladder safely, quickly, and efficiently, helping to ensure mission success. This training philosophy targets the lowest level of the organization, or what I refer to as the *Gumby* level.

The Gumby is the man going in the door first—the man making life-and-death decisions. He should be a primary focus of the training. He is on the tip of the spear and is responsible for the outcome of the operation. If he shoots a hostage or a friendly officer by mistake, it can mean mission failure for the organization as the rest of the unit will be all painted by the same brush of failure. It is our job as leaders and professional trainers to ensure his training and system is safe, simple, efficient, and lifesaving. Implementing complex, confusing, and unrealistic systems of shooting and CQB techniques will not help him in his journey toward consistency.

As a trainer and a leader, you should look at simplifying the process at every available opportunity. This should begin with tactics and missions. Generally, law enforcement tactical teams are assigned the big four:

- Hostage rescue
- High-risk warrant
- Search warrant
- Barricaded Person

Using dynamic and slow clearing techniques adds more to your training plate, and you will soon realize that you have a full one.

The key to developing simple tactics for all these missions is *not to*. You should instead focus on developing simple drills or a system that will apply to all the missions with minor changes. The reason is simple: Trying to develop a different set of drills that will apply to each specific mission is too much information for the individual to try to process. Further, you will never have the training time needed to become proficient at these drills. To add, the drills should not be directed at the experienced TL, but rather at the newest and lowest member of the team. Trying to memorize too many drills will only confuse and complicate matters, resulting in inefficient and bastardized techniques. Keep it simple.

I like to ask tactical officers in class, What is the difference between getting shot at on a hostage rescue, a high-risk warrant, a search warrant, or a barricaded person mission? The answer: none. They are all the same—you're still getting shot at. So why have four or five drills when one or two will solve all your problems on all missions?

A good trainer will come in and help keep it simple. A quality trainer should be able to give you a simple drill or two to handle all the contingencies for a particular problem. The key is to break down the training into blocks or modules. For example, I break down targets into five distinct phases for training and work on them as individual modules at first. They are as follows:

- Movement to the breach point
- Breach points
- Hallways and T intersections
- Close-quarter battle (CQB)
- Consolidation and reorganization

Teams first work on developing drills to handle all the contingencies that can come up during one module. Using the *movement to the breach point* example, I teach a drill that will allow the team to handle the following problems:

- Runner
- Complaint with a gun
- Suspect verbally challenging with a gun on the ground
- Shooter
- Shooter with an officer down

I first talk about the problem of moving to a breach point and all the possible contingencies. I then show them a drill that will safely address all the above problems. I allow the team an hour or two to conduct dry runs with a role player and get the movement and mechanics worked out. I then have the team load up with simunitions, and I pre-stage them around a corner. I brief my role player on which scenario I want him to play out. The team is signaled, and the role player acts out his role.

During one scenario, the role player will run, in another they will just shoot, in another they will be a compliant person with a gun. Each run is recorded on video, and five scenarios are run back-to-back. A short after-action follows after each run, and major points are addressed.

The film is reviewed in a classroom environment the next morning when everyone is fresh. Everyone looks at it at normal speed, and then if something catches our eye, we rewind the video and slow it down. We pick it apart for safety, tactical sanity if you will, and the effectiveness of massing fires on the bad guy.

We also ask the role player for his input as to what he saw. Were they mentally and visually overwhelmed? Was the amount and accuracy of the fire sufficient to neutralize them? Once we've looked at all the video and hashed out all the questions or problems, and if the tactic is agreed upon, we go do it again and reinforce it. Taking it a step farther, once the drill is mastered with simunitions or paintball, it can then be practiced on a flat live-fire range under controlled supervision. Team leaders can then run their team through dry, and then by giving one person on the team ammunition and checking the hits after each run. You can build up to everyone live-firing once the team has demonstrated proper safety and accuracy habits.

I believe that certain skills and training should be deferred to the individual level for maintenance and should not need supervision during performance. Shooting and physical training are two of these. With availability of weapons, ammunition, and ranges, individuals should take their own initiative to practice their shooting skills. As a TL, I do not want to waste my precious one day a month of training standing over my guys on a flat range. They should be able to dry-fire at night at home in their garage and then live-fire on their own time. I will spot-check them with team drills every few months and conduct biannual evaluations. The same goes for physical training. This is an individual responsibility, and we may go for a team run before training on training days just to keep everyone honest. On collective training days, team drills and collective training are the priority.

## TRAINING RESPONSIBILITIES AND SEQUENCE

Leadership at the team and organizational level must be knowledgeable about safety, tactics, and techniques to ensure training and mission success. Too many times our leaders are required to be administrators and are not savvy as to what the troops are doing and why. This is evident by the number of training accidents that for which training officers are directly responsible. Routinely these leaders are unchained from their desks, torn from their computers, and designated as safety officers. They have little or no concept of the safety requirements of the tactical training they are hosting or what can happen.

In my three years as a law enforcement trainer, I have kept abreast of training accidents at the individual, team, and organizational levels. When I started my training business, more tactical law enforcement officers were killed in my state by a string of friendly-fire shootings or by accident than by bad guys. It was incredible.

The typical organizational solution to the problem was to spend more money on training equipment—i.e., simunitions weapons—instead of taking a hard look at safety and training practices. This number has significantly decreased, but from time to time, you hear of officers screwing around, resulting in an exceptional officer or friend tragically shot and killed. This does not need to happen, yet it does.

As a soldier or law enforcement officer, you have the choice to be good at what you do or to be lucky. I choose to be good. If you choose to be lucky (sloppy), the law of averages will get you. And it will not get only you. The entire team and department will suffer.

A recent friendly-fire incident on a major tactical team during a real-life situation resulted in the loss of five officers. The officer that was shot was permanently crippled and is now working in an administrative division. The officer that fired the shot and the TL who was in charge of the team are now also off the tactical team. Two others left as part of their action or inaction during the incident. Five men suffered, including an outstanding career and personal life destroyed for the officer who was shot. This all resulted from violating one simple safety rule. I will not even mention the morale and litigation problems that will arise from this. It is a waste. Where can we point the blame? The individual, team, and organization are responsible.

At the individual level, the officer firing the shot violated a basic safety rule: *Lead shooter has priority of shot.* This means that if you're in the back of the stack or have officers in front of you, you can't shoot. You have to move in line with those officers, or in front of them, to safely make the required shot. You will get lucky four out of five times and get away with it, but the fifth time, you will tag your partner. As an ex-action guy, I would always be aware of soldiers around me, their weapons, and safety posture, because they could kill me just as dead as a bad guy. If someone did something unsafe, I would correct them on the spot.

At the team level, the TL needs to ensure that the team was trained up to the required levels to successfully complete the mission. This is where the TL needs to monitor team drills and ensure that everyone is adhering to the safety rules while in training. The TL also has to have the guts to correct team members when he sees them violating safety rules. Failure to do so will ingrain in them poor habits that will translate into live operations. Taking it a step farther to the organizational level, these leaders did not spot-check training to ensure that the TLs and members were following the proper safety guidelines and enforcing proper safety practices.

As a leader, the choice is yours. Invest in training with quality trainers, or poison your well. What would you want to see as a new student, a combat-proven leader or a "leftover" or screw-up who has no experience? Our business is too dangerous not to pass the right knowledge down through the proper instructors. Amateurs will only create an unsafe working environment. Further, ensure that your individual training complements your team training. Ensure that your team training complements the organizational training. Hold TLs accountable for both. Give them the time, assets, and personnel to do the job and then hold them accountable.

## TRAINING PITFALLS

Awareness of a trap is the first step in avoidance. With that thought, failure to maximize training time is a common complaint and issue in the tactical field. It is critical to use your training time efficiently. I see teams that have one or two training days a month. But the key to using this allocated time is to ensure your training day is planned, rehearsed, and ready to go with trainers, equipment, and locations. Some teams fall into the trap of the social training day. They come late, take a long lunch, and then leave early, feeling that they have accomplished their goals. They are sadly mistaken.

Units that fall short will eventually be compromised on a mission and will wonder what happened. If you do not take the training seriously or are not willing to commit time and resources, you're probably better off disbanding your team until you do. Some tactical teams' members are of the mind that unless they are getting the time either comped or off, they will not go to training. They feel that by simply being issued the special gear, they are qualified to do the mission. An old instructor said to me, "It takes five minutes to dress like a commando, but years to become one." This attitude of the "the gear makes me" is false. *You always make the gear.*

Another pitfall is the TL who sends his tactical personnel to every bit of training that comes up, but on their return, they fail to develop any consistent tactics or SOPs. Generally, tactical officers are the type A personalities who love to sharp-shoot and argue with each other. When officers attend a training course, generally they can absorb only 50 percent of the material that they are exposed to. Most often they miss the finer points. When they complete the course and bring this information back, it becomes a catfight of sorts to relay this information to the team. Old members will say, "That's bullshit, we tried that ten years ago." You find that tactics are like assholes—everyone has one, and they want to show it to you.

Generally, unless it is an open-minded team and TL, much of the information is suppressed, and the technique is not tried. Or the common response is that they will talk about it for a month, and bring it up at the next training meeting. People forget, or they move on to a new block of training and never do get back to the previous one.

Other teams have vocal dominant members who come back from training pushing a technique that they have just learned, and it becomes law. This is done even though they have never used it with simunitions, video, or in live-fire to verify if it is valid. Some leaders implement the latest fad technique without properly researching it. These new techniques will generally work dry, but throw in role players and simunitions, and the problems jump out at you.

Still others implement the latest and greatest technique just because the instructor was from this or that organization. This can be dangerous for several reasons. First, there are a slew of instructors who are one-hit wonders out in the community. One type of instructor is the trainer who has been to all the schools, and their walls are covered by certificates, but they have never kicked in a door on an actual mission. The problem here is the schoolbook answer versus the reality-tried-and-tested answer. As a friend once said, an instructor who has done only homework with no fieldwork will make a questionable instructor. Schoolwork must be accompanied by fieldwork.

Another type of instructor is the one who went to a tactical unit for a short time and was purged shortly thereafter, but because he was there, he now wears the badge of authority because he spent a year in that unit while on probation the entire time. He may be hailed as a subject-matter expert. Many times an individual like this was flushed out for a reason such as an injury, technical incompetence, or a consistent personality conflict.

I can deal with instructors who left because of injuries. Special Operations and law enforcement tactical operations is a fast-moving game, where you can easily be injured in training or on the job during a mission. Instead of pushing these types of individuals out of the arena, we should use them and glean as much information as we can from them. The military can easily put them into training committees, and law enforcement can assign them to training academies. The individuals in question may hate the idea of this, but I hate to see their talent and knowledge leave the organization. Training groups and academies are a great place to rehab as they routinely have an eight-to-five schedule, and gym facilities are usually available.

## REHEARSALS

Rehearsals are the lifeblood of a successful mission. This is not a fancy statement, but it is simple and true. As a private in the army, I honestly could not see the big picture of why rehearsals were important until I went to Special Operations. The reason was that I was at the Gumby level and not well informed. I was loaded down with equipment and told to march here or there and take instructions once I arrived. The other reason for my lack of understanding was that we generally did poor rehearsals.

As I would learn later in life, you need to maximize your rehearsal time, whether they be dry or wet, because your life depends on it. Dry rehearsals are generally conducted without live ammunition, whereas wet rehearsals usually include live-fire/ammunition training.

We discussed returning from a school or course with new tactical drills. These drills need to be validated before they are accepted as gospel. How do we accomplish this? We use all the tools in our training arsenal. Tools such as simunition, air soft, or paintball weapons can provide valuable feedback. We now have the ability to replicate getting shot at, and this adds a mathematical probability to our training. The probability deals with a ratio of hits from the bad guy while performing a certain drill or movement. Simply run your battle or tactic with a live role player who will shoot back.

Run the same drill several times back-to-back and record the training. Between runs, count how many hits the bad guy was able to deliver and how many the good guys were able to shoot. Keep a record of this, and then try another drill or tactic and keep the same statistics. Now review the video. First, look for a more aggressive system that masses your fires or lets everyone safely shoot. The tactic should also get everyone out of the kill zone during the contact. And finally, the drill or tactics should be safe and easy to remember. While reviewing the video, check the real time speed of the drill and determine which drill gives you a higher hit ratio than the bad guy. Your tactic should give you a 4:1 or 5:1 hit ratio to be considered valid.

Yes, officers can get shot, and will. This is the nature of the beast. This is also why we wear all the Kevlar and protective gear and carry guns. Further, I will tell you out of experience that the more bullets you can put into a bad guy, the faster they die. So whenever possible, safely mass your fire and make the threats go away sooner.

The added benefit of the training is the mental conditioning it will enable the individual to develop. Your personnel should train to bring all the critical marksmanship skills to bear at one critical moment, seeing your front sight on the target and squeeze. This simple act is what I term a fine motor skill, and it must be ingrained in your psyche. Force-on-force training will help you accomplish it when you train to do it. Using the above weapon systems, you can hold your personnel accountable for accuracy and lost rounds. You can also see where these lost rounds go and understand why it is important to make surgical shots.

In addition, students can now mentally condition themselves to fire and continue to fight, even though they have been hit. This important mind-set of fighting through became very obvious to me during intense combat operations. I have watched soldiers and leaders who were hit, with minor non-life-threatening wounds, mentally shut down, and give up while they still had the means to fight (see "Individualized Mentality: Fight-Through Mind-set" in chapter 1). This was not their fault, but rather, a training failure.

While moving through my service career, it was customary that while conducting battle drills, when you got hit, you stop. At the time, MILES gear was the rage, and when your nodule recorded a hit with a laser and your beeper went off, you were required to stop in your tracks. This only compounded and reinforced the mental "quit when you're hit" attitude. We conditioned a generation and an army to stop when they're hit, no matter the superficiality of the wound. I witnessed numerous soldiers who could have stayed in the fight, but instead mentally shut down and quit because of their prior erroneous training.

I inform students of this in my current training and tell them not to go down or quit unless I put them down. I advise students that even the simunition or paintball round strikes them right between the eyes, they shouldn't quit. Wipe it away and fight through—neutralize the enemy. Start the mental programming necessary for their survival during their current training. If I do have students that mentally give up, I require the team to carry them off-target as a punishment. Generally, students only give up one time, and their team talks to them about their mind-set as they are hauled out of the building.

This brings to mind another law enforcement raid where the officers were lucky that the bad guy did not have a fight through mind-set. This tactical team was forced by an apartment being located upstairs to enter a single breach point. This is a dangerous situation. It forces all the officers into one cone of fire for a brief second or two before flooding in and clearing the threat area.

The assault was initiated by a distraction device which was detonated on a "bang pole" outside the apartment bedroom window. The entry team entered the front door and rapidly took an armed bad guy down, who was sitting in a chair in the living room area. They then proceeded to the bedroom area to finish securing the apartment. Well, the other bad guy was in the bedroom with his girlfriend when the device went off, and he positioned himself behind the bed, cowering by the door with a 1911 .45 pistol.

As officers entered the room, he began shooting. He struck the number one man twice, once in the wrist and once in the head. The bullet went through the officer's goggles, entered his head above his eye. The bullet then traveled under his scalp and around his head, exiting to the back. The officer, knowing he was hit, called it out. His partner grabbed him and dragged him into a nearby bathroom. A third officer entered the room and found that the bad guy had fired the seven rounds in the magazine, and now he threw down the weapon on the bed and put his hands up, using his girlfriend as a shield. He was then taken into custody.

The problem I have with this is that the threat was not neutralized. Had the offender had a high-capacity magazine or a rifle with a thirty-round magazine, the story might be much different. The officer who was shot later recalled that it was like getting hit with a sledgehammer. The action of his partner—rendering aid first—was a conditioned response. I have always believed in the opposite: One dead body does not justify two.

You must learn to ignore the dead and injured and concentrate on neutralizing the threat or you will have more dead and wounded. These officers were fortunate, and the situation could have resulted in a more serious encounter.

## AFTER-ACTION REPORT/REVIEW (AARs)

AARs are critical to extracting lessons learned and ensuring that honesty, integrity, and candor are always held to a high standard within the force. Two types of AARs that come to mind are the informal debriefings that occur after a mission with either the key leaders or the entire group. The other is the formal AAR that consists of detailed written lessons learned that are passed through the chain of command for their information and to other units that may be going into harm's way. These units can use well-written AARs to adjust their training or to develop new and innovative training plans to deal with the problems or situations that were encountered.

Routinely, the informal AAR process is used during training and rehearsals to quickly identify problem areas and develop solutions. For example, your element is tasked with a mission to raid a walled compound in a small village. You develop a plan and find another small empty village about the same size and shape to do your rehearsals on. You begin your rehearsals with your company consisting of three maneuver platoons and a headquarters platoon. You task one platoon to block or seal the compound while the other two platoons assault. You run your dry rehearsal without role players on the target and go through your entire plan, movement, actions on and during the assault, consolidation, and then exfiltration from the target. All your platoons report back to your assembly or staging. You need now to elicit information that will be critical to make the next rehearsal even smoother.

You can gather every person in the company together, or you can bring in key leaders to the AAR. Since you're probably going to do further rehearsals, the troops probably need to clean and adjust gear, further prepare their vehicles, load up ammo, etc. While they are doing this, the key leaders can be sorting out what happened and how to fix it.

First, I would ensure that all squad leaders, platoon sergeants, and platoon leaders are present. I would then start by briefing on the overall concept of the mission and then ask for the key element leaders to describe what they saw, what problems they encountered, and what solution they have for fixing it. The key to successful briefings is to keep them professional, sterile, and not allow finger pointing. If someone screws up, it will be obvious to everyone. You want to create an atmosphere where elements are encouraged to bring up the problems that they encountered and how to fix them, so others may not run into the same problem. Fix it one time, the first time.

Generally, I will start with my lead elements and have those people brief on major issues and not delve into trivial matters. I really don't want to hear about who did what on their squad, unless it had a great impact on the mission. I will leave the squad issues to the squad leader to deal with. I want to know about how the mission flowed and if we can work on the big issues such as the sequence of our infiltration of the target.

Assault squad leaders might bring up the issue of a time delay on getting in and on the target with the way the formation is organized. They may suggest that if we send in the assault force to the compound first and then put the blocking positions in second, critical surprise time will be gained when penetrating the target. This is a good point, and if enough of the personnel agree that this is a better solution, we change it. I prefer to let the individuals that are kicking in the door to plan the plan and execute the plan. As a commander, I am there as a safety valve or sanity check. I will look over their assault plan and ensure that we do not have any fields of fire problems or doing something stupid that will get soldiers killed.

Another technique for debriefings is to keep leaders from rambling. This can be done by reviewing two positive points and two areas that need to be improved. This keeps things simple, and as a commander, I would focus on fixing two to three things with each rehearsal, and this will help in the planning and rehearsals exponentially. If you have too many key leaders trying to talk or you have the ones that like to hear themselves talk, you will bring out too many trivial issues, and you will forget what is important and what is not. Keep it simple, controlled, and nonpersonal. If individuals have personal issues with other leaders, let them take it up after the briefing. The key to the informal AAR is to get the information disseminated and back out to the action guys so they can do another rehearsal.

The first informal AAR may take some time if your plan is rough. But as you fix the two to three items with each rehearsal, your actions will smooth out, and the AARs will shorten to probably just a few minutes with little or no input. Again, it is critical for the guys on the ground to get as much time working through their actions as possible.

## MAINTENANCE TRAINING

Maintenance training for a force can take two forms. The first is a yearly check of the block training that encompasses all the missions that the unit is required by their charter to perform. The other is when a unit is deployed into a threat area and you have not conducted a mission of any type in some time, generally a week or so. These refresher missions will help keep the unit sharp and focused should the call come.

Routinely I read the chat lines where tactical teams complain of being burned out and not having anything to train on. I just shake my head (see chapter 18).

## EVALUATIONS

When I walk into an unfamiliar organization to do an evaluation, I keep my evaluation system simple. I am professional, polite; I keep my mouth shut and my eyes open. What am I looking for? Everything.

First, I look to see how individuals perform at their level and what their motivation and initiative is like. Next, I look to see if they do things on their own initiative or if they have to wait for guidance from above. This will readily indicate the type of leadership style present in the organization. Hopefully, there is a sense of empowerment to the lowest level. If not, it will probably be a tightly controlled authoritarian or near dictatorship. I look at their individual skills, such as weapon safety and how they handle their firearms and their load, unload, and clearing procedures. Is it safe, do they have a system for safety, and are they conscious of it?

I then watch the TLs and see how they are performing. Do they spot-check and pass information as it becomes available? Does the team have a scripted system of doing business, or does the TL have to point out every detail to team members? Is there an ATL present ensuring accountability for the team and their equipment, freeing up the TL to begin planning?

Observing the command staff, I look to see if they are organized and efficient and whether information freely flows to the team level. Do leaders come in and take charge, or do they wait for key personnel, the "old reliable," to come in and make it happen? I watched this happen with a great organization on a large exercise one time. Key personnel that routinely took charge intentionally delayed their arrival to the crisis area to see what their subordinates would do. In this case, their subordinates were so conditioned to their leaders being there and giving direction and guidance that these subordinates waited, and waited and waited . . . All the while, the scenarios were still playing out on the target, as they do in real life. It was a great learning experience for everyone.

Back to training evaluations. Looking at the training and tactical scenarios that an organization sets up for their team will provide you with an accurate snapshot of the importance and professionalism of the entire command staff. If the staff puts a great deal of work and thought into scenarios, it is a good indicator that the organization takes the training seriously.

Routinely I am asked by individuals and units that I train, how do we stack up? To give an honest evaluation of a tactical unit, you must first know what is at both ends of the spectrum, so to speak. Using the yardstick analogy, I was able to see organizations from all over the world operate, and I would have to use my base of reference to determine their level of performance.

This may seem like a simple task, but it is dependent on several variables of the tactical organization. You must first look at the unit's mission or missions and determine if they have the number to do the job. Smaller departments sometimes have tactical teams whose number barely reaches into the double digits. Conducting a complex multi-breach-point hostage-rescue operation generally requires twice that number of personnel. If that team has not worked out a mutual aid agreement with another agency, they may lack the personnel to properly do the job. Of course, they may execute the mission anyway and be lucky, but we should not rely on luck in this business.

Next, available personnel for selection can be a critical component for a tactical team. If the host department only has twenty-five sworn personnel and ten are on the tactical team, you may wonder about their capabilities. On the other hand, I have witnessed departments with thousands of officers where the management did not care for tactical units and they deliberately kept them undermanned. What key components have a direct effect on both these issues? Leadership and conviction.

I have witnessed small departments that could not afford the overtime, but the guys on the team used their own time and money to ensure they had the proper gear, and they conducted enough training to sustain themselves and to properly execute a mission. This not only shows heart, but it also shows incredible conviction in the mission.

## KEY POINTS

- Train as you fight.
- Validate all tactics and drills prior to combat.
- Use videos whenever possible for AARs.
- Ensure AARs are honest and promote candor. They should focus on solving the problems and not pointing fingers.

# 8

# LEADERSHIP PLANNING

*Never tell people how to do things. Tell them what to do and they will surprise you with their ingenuity.*

—George Patton

- **YOUR PLANNING SYSTEM**
- **ONE OR MANY PLANNERS: WHAT THE SITUATION CALLS FOR**
- **REHEARSE YOUR PLANNING SESSIONS**
- **EMPOWER YOUR LEADERSHIP**
- **THE QUICKEST AND MOST EFFICIENT WAY TO PLAN**
- **CONCLUSION**

## SCENARIO: THE MISSION

We received another intel from a "reliable" source that our priority target was spotted in a vehicle and had been driven to a residential house. The TLs quickly dressed, secured their weapons, and reported to the TOC. The information was simple and straightforward. The "big" man on our target list was spotted by U.S. forces leaving a foreign embassy. The forces notified our assets, and eyes were assigned on the vehicle. The vehicle entered a walled compound that was located in an upscale portion of the city. We checked the area and found that there were two to three vehicles present in the compound.

As the forces assembled, a new AC was rotated in to perform the command and control responsibilities of the group leader. This particular leader had been on other hits and watched the planning process but had never actually performed the job before on a real mission.

As we gathered around the dry-erase board in the TOC, the commander grabbed a dry-erase pen and decided he was going to take charge and designate how the assault was going to go down. As TLs, we just sat back and watched.

He had been programmed through his leadership training to take charge from the front and lead. What he found as he watched our hasty planning session was that with the amount of assault, support, and cover aircraft in our package, there was too much information for him to process quickly. He looked at the sketch of the target for a moment and realized he was in over his head. He handed the dry-erase pen to a sergeant major and turned over the planning phase to the tactical element leaders.

The assault TLs looked at the sketch and let the lead team pick the point of insertion where we could get our four packages in. The assault pilots also looked at the plan with them to ensure the area was large enough to get the bird in. Then the security guys came up and picked out the best positions to set their blocking forces in with their pilots. The gun pilots sat back and watched both, figuring the best plan of attack to ensure that they could provide the best cover should we encounter threats. It was done, in two to three minutes, and we had a plan to integrate assault, cover, and attack aircraft into one simple template plan. We moved out to our birds, briefed our team members and our individual pilots to ensure we would be landing according to the plan. We then mounted up on our birds and waited for the liftoff signal.

This run would be a fast, almost-straight-line run into the target. Blades were turning, and we began pushing air and started our move. The initial snakelike formation was inbound, hot and moving with a purpose. We got the one-minute from the pilot and unhooked our safety lines. The attack birds did not pick up any threats and screamed past their targets. We were hot on their ass. We were scanning hard on our final approach, and our pilot picked up a wire in our path at the last minute and had to do a near-vertical insertion. He smoothly touched down, and we broke to the right side of the bird.

We faced a compound wall with two heavy metal gates to our front. On our left was a small corner snack stand that jutted out off the main wall, one to two meters deep. We ported the window with our weapons to ensure it was clear of potential threats and refocused our attention on the gates. Team breachers were already placing charges on the gates, and we were looking for places to hide from the impending blast. They gave the blast signal "fire in the hole," and we braced for the shock of the charge.

Heavy metal gates required a somewhat larger charge with the old military P for plenty of explosives factored in. The charges went off, and it rocked us a bit, but we were used to it by now. We would get ninety degrees off the blast plane and keep our mouths open so as not to overpressure our lungs. We were probably only fifteen to twenty feet from the point of detonation, but the ninety-degree factor and the heavy walls helped protect us.

The sharp crack coupled with the dust clouds and debris that blew past us let us know the charges had gone off. We moved in and entered the gate to the right, as the gate on the left was not yet open. We moved down a long driveway in a modified wedge and focused on the house to our right. Remembering what the target looked like in the planning session, I knew we were at the wrong house. Our photos showed two mirror houses, one on the right and one on the left. We were approaching the right one, and our target was over the wall to our left. Looking left, I saw a doorway in the compound wall and knew it had to lead to the right house. I told the team to move left.

As we punched through the gate, we crossed a short patio and entered a dining room with a table full of hot spaghetti fit for a king. We started to clear and began to secure everyone. We quickly took the dining room and kitchen down and then moved down the hall to the living areas. A thirty-foot hallway with two doors right and a door left ended with a far door that led outside. We put a woman down in the hallway coming out of a room on the left. We cleared the first room on our right, and then a door slammed shut on next room down. It made our hair stand up as we pushed down the hallway. Focusing on the slammed door on our right, we covered it while we cleared the room the woman had come out of. When the team gave a verbal clear, I told my breacher to "charge it."

I grabbed the woman who was lying on the floor by the hand and dragged her back down the hallway around a corner where she would be safe from the blast. I moved down the hall and took up a position in the room across from the charge. My breacher called "burning" and pulled back into our room. The charge went off, and we moved through the smoke and cleared the room. It was empty, and it appeared that the wind had blown the door shut. We came back out and cleared through the door at the end of the hallway, only to link up with a team who was pinning the rear of the compound.

Moving back in, we started to consolidate all the personnel on the target into the living room area. The men were placed on one side, women on the other. Looking over the men, I knew what had happened. We had a look-alike—a person that looked like our target, but this guy was much older and fatter. Our intel sources had mistaken this person for our target, and we thought it was better to take action than to kick ourselves later for not taking action. We handed over our prisoners to the teams designated to further search, control, and move them. Reporting to the AC that I had all my personnel, we prepared for exfiltration.

### AFTER-ACTION COMMENTS

*SUSTAIN*

- Stay alert and stay in the fight. If you're not where you are supposed to be, move to or fight your way to where you need to be. Always look and assess.

*IMPROVE*

- If it is not broke, don't fix it. If the planning sessions you are using are successful and work, keep using them.

---

## YOUR PLANNING SYSTEM

The planning system you choose should be generic, flexible, and fast. It should incorporate all the talent and leadership assets at your disposal. The same system should be practiced and rehearsed during all your training sessions to enable you and your subunit leaders to refine and streamline their portion of the planning process. Further, the planning staff should be geared to support you and your mission and not delegate policy or tactical guidance to you. Their job is to facilitate the information flow and tactical assets to you and your men. This information should be both verbal and posted.

## ONE OR MANY PLANNERS: WHAT THE SITUATION CALLS FOR

I have seen just about every type of rapid or deliberate planning session throughout my military and civilian career, and the one that I have seen work the fastest and most efficient in a high-stress environment is *collective planning*. This session is where all tactical, maneuver, and support leaders are present to ensure the plan is understood and well represented. These are the people who are going to kick the doors (assaulters) in, the people who are going to get them close to the doors (pilots/drivers) and the people who are going to protect them from the outside (support) once they are in the target.

Who begins the planning process? The commander should immediately start by giving an overall mission statement, then hand the process over to the elements who will next determine the priority of infiltration. For example, if you intend to put your assaulter on the target first, the assault TLs should take the lead in developing the insertion plan, with the pilots or drivers right by their sides to ensure that you can get the air assets or vehicles into where the TL wants them. The drivers or pilots may say their vehicle will not fit and cause you to adjust your plan. You want this to take place at the planning board and not on the ground or on final approach while you are taking fire.

Once the assaulters get their plan down, the security force then needs to figure out where they can best protect and serve you. They should conduct planning with their delivery personnel, either pilots or drivers. Assaulters should be looking at this area and ensure they know where their protection is for both their fields of fire or incoming hostile fire. For example, putting a security position too close to you will cause you a headache. When the enemy engages the security position, the rounds that miss the security element will bleed over into your position. Ensure you have cover between you and your security elements.

Once everybody understands the plan and there are no more questions or changes, you move out and brief your men. Using a simple sketch map, you point out where you are going, and if you can't get into your primary infiltration point, know where your alternate is and how to get there. The pilots or drivers must also confirm these locations. If there are no questions, you are ready to do business.

## REHEARSE YOUR PLANNING SESSIONS

Planning staffs should be required to be at routine training to ensure they have a system that best supports the men going into harm's way. The planning and briefing area should be set up as it appears in combat to ensure the most efficient system has been implemented. The staff and intel personnel should be present to feed information to the end user in a quick and efficient manner, as they would in a real combat mission. Only by requiring them to participate in routine training will you ensure that they are able to perform these critical tasks.

Many times commanders allow the requirement of the day-to-day unit business as an excuse not to attend training. This excuse may fly once in a while, but commanders should ensure they are there and their system is in place. Remember, these intel and planning personnel are generally working in a warm and dry environment when the action guys are freezing, baking, or getting pissed on with rain. The least they can do is to work a few more hours in a controlled environment to ensure their area of responsibility is functionally intact.

Rehearsals should include the entire process of alert and notification for a mission as well as deployment of an advance team if required to include a reception of the main force. The planning process should then integrate the entire package through the actual assault or follow-on assaults. This may be a one, two-, or three-day process. Afterward AARs should be conducted to ensure that the process worked and the men in the arena had all the support and intelligence they needed to accomplish the mission.

Overseas, simple coordination for latrine facilities for over twenty-five personnel can be a major problem. Sanitary conditions need to be maintained to ensure the force remains well and healthy to accomplish the mission. A parasite will put a force out of action as rapidly as an enemy bullet. For extended operations, sleeping arrangements need to be worked out to house the forces and allow them proper rest and not expose them to the constant commotion that comes from a TOC. This may be solved by a simple issue of eye patches and ear plugs.

## EMPOWER YOUR LEADERSHIP

Empowerment is critical to ensure mission success. Your subunit leaders must have responsibility and understand that it is up to them to make it happen. This includes using their tactical thoughts and input and ensuring they are always part of the planning process. This empowerment accomplishes several things. First, it requires each subunit leader to take initiative and have their personal system squared away and ready for the fight. While these subunit leaders are planning, usually their troops have the time to tailor their equipment for the fight. The planning leader will have little time to do this because of the planning requirement. If they have a cohesive team, the team will usually take care of it for them.

Empowerment also puts the mental and physical responsibility of the team on the TL's mind. This creates an added positive pressure to exceed standards when one knows that other lives are at risk. They subtly know that their skill at planning is critical to ensure the survival of the team and accomplishment of the mission. Leaders at this level should ensure they are always mentally and physically ready for the challenge and that their teams are up for it. To ensure that your team is ready to conduct rehearsals, rapid and accurate information must be passed down to them in a timely manner. This can be accomplished in two ways. First, the TLs each go directly to their prospective teams and relay the information. If planning is going on, one TL can go and inform a member of each team. A second technique is to have one ATL report to the planning area, brief him, and then have him deliver the information to the rest of the teams. The information pushed down to the ATL could be as simple as an additional equipment requirement for the mission or something as critical as information on the number of bad guys and hostages.

Further, TLs who plan together will be more apt to work with other TLs regarding the critical portions of the mission such as alternate breach points and link-up points inside or outside of the target. Alternate breach points are always planned, and should a team not be able to get into their primary breach point, the TL will direct them to the alternate. When moving to an alternate breach point, a team will be pushing through and into another team's area, coming in behind them to link up, either moving through the team to their front or pushing them ahead farther into their area of responsibility.

Link-up points are another critical area that TLs need to plan and rehearse. A link-up point is a physical location inside the target where two friendly forces will find themselves pointing guns at each other. Sometimes a bad guy runs from one team area and into another team, and the gunfight starts. Teams need to know when to push and when to hold back and when to push through should you wait on a link-up team too long. In past incidences, link-up teams have mistakenly fired on each other, mistaking a member of their sister element for being a threat.

## THE QUICKEST AND MOST EFFICIENT WAY TO PLAN

### *Alert the Force*

Ensure that the system you use to alert and gather the force works and is simple and effective. Have a primary and secondary method of communication available.

### *Report*

All members should report in, and leaders should be given a brief mission statement that can be passed on to all team members. If an ATL gets in first, he can pass the info to the TL upon their arrival. If the TL is late, the ATL should begin the planning process.

### *Begin Planning*

After briefing their teams, TLs should report to a central planning area to receive an information update and command guidance. Other key things happen at this time, but this being an open source of information, I will not elaborate on special team tactics and deployments.

### *Pass Information On and Prepare Rehearsal Areas*

As information becomes available, it should be passed down to the end users (teams) as rapidly as possible so they can tailor their gear for the mission. Also, certain teams may be responsible for getting together a rehearsal site, and this can be as elaborate as doing a live-fire scenario in a local shooting house or as simple as a walk-through tape drill on the ground.

### *Brief the Plan*

Once the plan is formulated and the commander gives the go-ahead available assets should be assembled and briefed as a group, if at all possible. This technique involves everyone looking at the plan, not to sharp-shoot it but to look at common danger and problem areas that need to be addressed. So when you talk about link-ups, you understand which team you are going to see.

### *Rehearse*

Rehearsals should be conducted from the bottom level up and then the actions on the objective first. Hopefully, once ATLs get the word of the mission and get their gear ready, they should have taken their individual teams and started to do walk-through rehearsals on their own. They can walk through and discuss exterior movement, breach point procedures, interior movement, CQB procedures, and consolidation procedures for the target. They can also review medical procedures and do an equipment shakedown to ensure all the mission's essential gear is present and functioning. Much of this concerns individual and team actions.

### *Fix Problems*

Rehearsals should start with all teams working on how they are going to take down the target, or actions on the objective. They should do this several times to get it done right, and then the rehearsal staff should throw in some problems to make the force think and ensure their backup plans work and are viable. Also, rehearsals will point out any glaring problems with the plan. Once this is accomplished, you can then work on your movement phase and on putting the entire package together.

### *Re-Rehearse, Re-Brief, and Integrate All Assets into the Plan*

If there are too many problems with the initial plan, leadership should re-brief the entire plan to the group. Too many changes will confuse everyone, including key leaders, as to what the current plan is. At the individual level, men need to have a simple and streamlined thought process as to what is going to happen on the mission and their role in it. Taking the time to re-brief will save everyone a world of headaches.

### *Prepare for Mission Execution*

Once final rehearsals are complete, the force can catch their breath and stand by for the execution phase. It may come soon, or the action may take days. Leaders should act accordingly and ensure that the men are not burned out by leaning on the edge too long and too hard. Take a down time as appropriate and ensure that the men are rotated if it looks like they are becoming fatigued or burned-out. Then re-rehearse actions as needed to keep the force sharp.

## CONCLUSION

Leadership planning is a critical component of an operation and must be rehearsed and refined as any other action. Further, junior leadership should be continually exposed to this process to cultivate future leaders and shorten their learning curve as it relates to their leadership growth. Leadership training will be covered in the next chapter.

## KEY POINTS

- Develop a simple and effective planning system and then rehearse it.
- Empower and use all available subunit leaders and their talent to develop a plan.
- Rehearse the plan, and if there are too many changes, rehearse the entire plan.

# 9

# COUNSELING AND MENTORSHIP

*A neglected skill*

- **COUNSELING**
- **THE IMPORTANCE OF THE OUTSIDE PERSPECTIVE**
- **ADMINISTRATIVE REQUIREMENTS**
- **TOTALITY OF AN INCIDENT**
- **TECHNIQUES AND COURSES OF ACTION**

## SCENARIO: REMEMBER YOUR FRONT SIGHT

"Remember your front sight," was probably the most memorable and lasting statement a leader has ever said to me before I went into combat. It came from our group sergeant major the night we were going to execute combat, operations in a country that was finishing its downward spiral. He stated, "Tomorrow night you will be in combat." A few days earlier, this country had crossed the point of no return: Its security forces had killed an American officer. The U.S. military and political leadership finally had the reason they needed to conduct a large-scale military operation and remove the country's corrupt government, one that we probably had some help in establishing.

We loaded our gear and flew into the country in question. We retrieved our cots and set them up as we had done a couple of times before. We then waited. At the Gumby level, I carried the M203 grenade launcher and a few extra rounds. The tropical heat was playing hell on our bodies, and acclimating was a bitch. The heat caused our bodies to pour water out.

Members of another group tried to wear their heavy Kevlar vests and conduct training, but the heat set them straight. I watched as more than one man came back from a training exercise and collapsed on the hangar floor from heat exhaustion. Soon they would be hooked up to a 1000 ml IV bag, and once it was gone, they would probably get one more. The rule of thumb I learned during medical classes was that when you had to piss, your body was starting to get enough fluid back.

We found that we could not physically put enough water back into our bodies to make up for the loss caused by the body armor we wore at the time. The armor, coupled with the constant heat, would put us of action as quickly as an enemy's bullet. The leadership had a real dilemma to deal with in this situation. We were tasked to perform multiple operations, either day or night, and this tasking said nothing about IVs being plugged into us while we were conducting these missions. The risk vesus gamble option was put out, and the decision was made not to wear our Kevlar vests. Instead, we would generally wear a pistol belt and a chest pack or load carrying equipment/load bearing equipment vest (LCE/LBE) of our choosing. In the end, the individual had to be able to hump his load mission after mission and be ready to perform for the next one. I was shocked that the chain of command had let us make that decision, but I was also impressed. The group commander was a no-nonsense guy who was not afraid to make smart decisions or let subordinate leaders make wise decisions. He did not look at his career first, as so many did at that time.

The combat operations, or live-fire ARTEP as I referred to it, (Army Training Exercise), was kicked off on schedule; and our sister element led in on the initial wave. They were successful in their hit and returned with minor casualties. The overall commander, a consummate ticket-puncher, allowed us out the next day to go to work. We were able to flush the elusive target after a week of beating the bush. We conducted over forty hits in a week, keeping the bad guy moving and guessing. While conventional units were trying to push through the city on a dedicated time line, we were able to roam free from one side of the city to the other.

As intel came in, we hit targets in all areas, keeping the momentum and presence of U.S. forces intact. We would jump on a neighborhood and rumble in, shaking the ground with our approach, and jump on a target of our choosing, wreaking havoc for a short time until we were sure that our desired person was not there. We would then move on to other areas of interest.

Generally, no one opposed us on our missions. On one instance, two guys committed suicide by shooting at us as we neared our target building. The vehicles we were on board opened up with at least four Browning M2 .50 caliber machine guns at what I still remember were the most beautiful cones of fire I had ever seen. The tracers could be seen in the night, flowing from all the guns into one spot and, upon impact, became unstable and erratic. To add to their firepower, we poured out of the vehicles like ants, launching a few tracer rounds or 40mm high-explosive grenade at the beaten zone. The two individuals, who were probably a local home defense force, made the deadly mistake of firing at our twelve-plus vehicles. They did this with Uzi submachine guns. Dumb. We were hunters on a mission and ready to do business. Their world quickly came to an end.

We had been running hard for three or four days, and we ran into these two jokers on a night operation. We had executed numerous raids to this point, and we were getting burned out, mentally and physically. When the lead bad guy opened fire, we pushed out from the vehicles in all directions and set up a loose perimeter. Our team went straight ahead and made a right turn and then pushed a little deeper down the street. We did not know where the target building was, but we were going to push our perimeter out far enough to protect the force.

We made contact with some locals who were the "home guard" of the next neighborhood thirty yards to our front at the next intersection. We jacked them up and put them on the ground. Searching them, we found a .38 revolver on one. I stuffed it into a cargo pocket on my pants. We flex-tied the man who had the gun and dumped him into a flower bed across the street where Tony and I pulled security down the intersecting street.

Gary and Al stayed on the main street to our front. All of a sudden, someone poked out of a darkened house and pumped six rounds down the street in front of us. No one returned fire. It sounded like someone with a revolver shot their wad and tucked back into their house, playing it safe. During this time, Tony and I saw folks moving around on top of the hill above the target building and vehicles. I held my fire as I did not see any gun, but suddenly a .50 cal from the vehicles sent a burst up, and I could follow the tracers to their target. I swagged it (best guess) and sent one 40mm from my M203 grenade launcher up into the area the tracers had just impacted. If nothing else, I hoped to help keep them honest and their heads down. We never did take any fire from that area. While covering Tony's six-o'clock, the guy we had laid in the flower bed called to us and asked if he could move. We laughed and told him, "No." The truth was that we had forgotten all about him.

We got the word to pull back, and things got hairy. We did not know where our other security positions were, so we bounded back, using cover until we linked up with our guys. They were switched on and recognized our movements, shapes, and outlines and held their fire. A few of the guys were holding on the other side of a vehicle that had been parked sideways in the street to prevent other vehicles from entering the neighborhood.

You could see vehicles all up and down the street set up in the same fashion. Before we moved to the opposite side of the car, I glanced down and saw someone lying in the street on the other side from our guys. Apparently, when the shooting started, he hit the ground and was too scared to make a noise. He stayed there the entire time. I told the team what they had, and they talked to him and told him it was OK to go back home. It seemed like we had only been on this target for an hour or two, searching buildings, consolidating the dead and wounded.

We pulled security by the vehicles and waited for the word to mount up. We were smoked, physically tired, and running on empty. I equate a firefight to that of a parachute jump, but worse in terms energy expended. Usually I tell folks that one parachute jump robs your body of about eight hours' worth of "juice" or energy. The same goes for a firefight or engagement.

We had been running for three or four days straight doing hits in the daytime heat, and it was catching up on us. Prior to and during a hit, you would get an adrenaline dump that would keep you pumped up for an hour or so, but then you would want to crash. Pulling security in a buddy team, you would tell your buddy to rack out for a few minutes while you would pull security, and then you would take your turn. You have to get rest whenever you can. We received the word to mount up, so we collected our team and remounted our vehicle.

We started our movement back toward the center of the city, where we ran into a conventional force that was on foot. We got the word that some of their privates were trying to put their antitank weapons into action and engage us. This is one of the things that unnerved me about combat. These folks had been getting sniped at and were understandably edgy. Even though we were traveling in U.S. vehicles, of which the bad guys had none, our soldiers still got scared and wanted to pull the trigger on this perceived threat.

Looking ragged with a three-day growth of beard, we were developing that hard look of a "don't fuck with me" appearance. A fire support captain attached to that unit came out of a dark doorway and looked at us and a couple of our vehicles and asked, "Who are you, guys?" Norm sounded off with, "We're the ghost busters." I laughed to myself and told the captain to talk with the officer on our lead vehicle. He kept looking at us, saying, "Nobody will give us any fire support. Spectre will not even drop illumination for us." I did not doubt why. No telling who this unit would fire up in their travels.

We got the word to hold fast, and two of the trail vehicles took off in the direction of our last target. We came to find out it was good that we stopped. During the stop, one of the guys on the rear tracks was looking at the back of a helmet of who he thought was his battle buddy named Kit. When they stopped and the guy turned around, it became apparent he was not the buddy. Evidently, during the lull in the last action, it was Kit's turn to nap, and he did. But when his battle buddy remounted their vehicle, he took the back of the helmet of another guy for Kit and did not see his face. Kit was still asleep on the ground at the raid site. Evidently, Kit woke up shortly after our departure, looking around and seeing nobody there. He thought, *What the fuck!* and got into some bushes. He then got his radio out and started to make calls. At about the same time, our guys realized that Kit was not in the vehicle. Kit got on the air and let them know his situation. The two trail vehicles went back and recovered him without incident.

AFTER-ACTION COMMENTS

*SUSTAIN*

- Maintain vigilance around friendly troops.

*IMPROVE*

- Get face-to-face accountability whenever possible.
- Ensure your battle buddy is accounted for.

---

## COUNSELING

Counseling is the process we use to correct behavior that does not fit into the desired scheme of things. Through counseling we advise or give suggestions to encourage positive behavior. I would say that this definition is darn close to that of leadership except that counseling occurs at the individual level. Counseling can go up or down, left or right. More commonly, it goes down to subordinates, but you can counsel peers in an informal way. This also applies to your boss, though it would probably behoove you to ensure your delivery method of counseling is respectful and professional.

The counseling of peers should, at a minimum, consist of three forms. These forms being initial, quarterly, and exit counseling. Initial counseling should consist of both a getting-to-know counseling and an orientation/information counseling. The getting-to-know counseling is just that, getting to know your new person at the individual level, taking a personal interest in them, their personal history, their family, and their future goals. This counseling should break down any preconceived barriers or walls established by the mystique of the organization. For example, an officer that is selected for a position with a new tactical team probably has a great deal of preconceived ideas as to the organization, mission, or structure of the team or unit. This would be a great time to dispel these myths and set the story straight. Another suggestion is to do this with your next-in-line subordinate, or at least during an introduction. Once the individual session is complete, you probably want to schedule a follow-up session for more of an orientation to the new organization.

A simple suggestion for a follow-up orientation is to give the new person a binder containing all the organization's rules and SOPs to look at and review for a week or so. Schedule the next orientation session so that the new person has time to read and ask questions about the rules and SOPs. At the next session, go over any highlights that you see as recurring problem areas, and then have the new member sign a memorandum of understanding as to having read the book. This will help dispel any future excuses such as "I didn't know" or "I wasn't told." It might even be wise to assign the new member a sponsor who will work closely with him for the next few weeks or months to get his feet on the ground. This battle buddy concept generally ensures that a team concept is present, and it will greatly strengthen the individual's new view of the organization and the team-oriented work ethic.

Once the individual has his feet on the ground, the unit leader should conduct quarterly or semiannual one-on-one counseling. This approach clears the air of any developing misconceptions and ensures that everyone is working toward a common goal. Occasionally, individuals have a perception of themselves that is not shared by leadership. You owe it to this person to let him know. Individuals may be totally unaware of what they are doing, not doing, or their current level of performance. It is the leader's job to look such individuals in the eye and let them know. This should be done in a tactful manner; and the more professional and sincere the individual, the smaller the hammer you will need.

There are some great people, but some are dumber than a box of rocks and require a much bigger hammer to get the points through to them. Some personalities are so confident and overbearing that you have to chop them down a notch just to get one or two points across. Evaluate your target, and then choose your delivery method.

One method to break the ice during counseling sessions is to let the person talk first and burn up their nervous energy, then you can ease into your counseling. Sometimes your counseling may consist of nothing more than saying you're doing a great job and keep it up. Either way, give the person a chance to get comfortable and establish the session as a positive one and not a one-sided ass chewing. When bringing out deficiencies, have your facts and figures straight and offer suggestions on how to improve. Ask the person what he intends to do to improve his weaknesses. By allowing him to feed you back information, you are ensuring that he understands what you said.

Exit counseling is used to help make the team or organization stronger. I generally do this once an employee is designated to leave an organization. Once all evaluations are complete, I want to ensure the individual understands that nothing will be held against him once he departs, nor will any evaluations be changed. Generally, this will produce an honest and informative session that can be used to better the organization. No matter the candor and honesty that is established in individual counseling sessions, occasionally, individuals are apprehensive about bringing out negative points about internal team issues, feeling that it is a team problem to work out. This technique will sometimes bring to light issues that you are unaware of and will keep you abreast of problems that might fester.

Spot reports, achievement awards, or "pulse checks" are other types of counseling that can take a formal or informal status.

Spot reports, or blue cards, were cards used in ROTC to denote a onetime good or bad incident involving a cadet. These spot reports were generally issued for negative conduct, but could as easily be issued for positive conduct. Too many times in life, we are only worried about bringing out the problems in a person rather than the good points. Good spot reports add ammunition and substance to counseling and yearly evaluations and reinforce positive behavior.

It can take about five minutes to write an evaluation—you can check the block, jot down notes of the incident, and talk to the person about it in a formal or informal setting. If it is a good incident, you can praise them at the individual level in front of their team or the group. Keeping the positive record and others like it will allow you to articulate it later on counseling reports and yearly evaluations as well as to pinpoint and substantiate higher performance levels given. Too many times we scramble for input at the last possible moment to substantiate an evaluation. In some cases, evaluations are downgraded because leaders do not do a good job of counseling or record keeping. The individual in question suffers because of this.

Negative spot reports are equally important. First, you should try to correct the behavior by giving the individual all the help and guidance he deserves. The rest is up to him. You may need all the incidents written down as reference should you need to permanently remove this individual.

When giving a negative spot report, talk to the individual as discreetly as possible after the incident. You can use a blue card if you choose to be formal, or if the incident is that severe, you can write a formal counseling statement on the action. Corrective counseling can take almost any form, from, "What were you thinking?" to "Let's go talk." Personnel in the tactical arena should be thick skinned and ready for direct and frank feedback. During rehearsals, this should be addressed in front of the team first to correct any misunderstandings and to ensure that they understand what the problem was. You can then take the problem individual off to the side for a talking-to. You have these options at your disposal for addressing the problem:

- Discuss off to the side at the individual level.
- Discuss in front of the group and let it be.
- Discuss in front of the group and jot it down for future reference.
- Discuss in front of the group and give a spot report later.
- Discuss in front of the group and formally counsel.
- Discuss in front of the group and remove from training.

If you think it is just an isolated incident, you can let it be. If you are not sure if it is an isolated incident, you can jot down the incident on a blue card and keep it for record. If it is serious enough, you can give the individual a spot report and counseling of the incident, and have him sign the spot report indicating he was counseled. For the next level of seriousness, give the person a formal counseling statement either the same day or the next day to ensure they know it was a serious infraction. If the infraction is a safety issue that cannot be tolerated, you can dismiss the candidate from training and have him report later for counseling or counsel immediately. The choice is yours. It is important to remember that you have many tools and levels of discretion to use when working with your people.

Achievement awards are simple tools that a leader can use to reward exceptional performance. Use them when you have a team player who exceeds the standards and consistently performs in an exceptional manner. Officers I have worked for have used the term *tools* to describe the use of awards. These pathetic officers are not real leaders. Routinely they would change the standard for the award to their own personal standard and sometimes made it a personality contest or a political tool. Awards are simple. If someone takes the time to do something great, you as a leader should take the time to see that they are rewarded for their effort.

## THE IMPORTANCE OF AN OUTSIDE PERSPECTIVE

Outside perspectives can be ascertained by pulling your ATL to the side and asking him his opinion on the person or incident in question. If he is too close to the problem, you could ask another TL his opinion on the matter. If you wish to involve upper leadership, you could ask your boss and get his two cents' worth, first ensuring that they will not take action without your recommendation. Most good leaders won't, but occasionally you will get one that may have a grudge against either the person or the action the person did and want to jump the gun and hammer the person instead of letting you use your tools to work it out.

I used video in training as another way to ensure an accurate picture of a team member's performance. Military and law enforcement conduct training is more hazardous than other types, since there is zero tolerance for safety mistakes or sloppiness. You may wish to target this area because an individual is having problems with decisions or is occasionally sloppy with a muzzle.

In Special Operations, most students were washed out of training for either safety violations, discrimination, or for CQB problems. CQB requires a constant focus on safety and discrimination. New candidates who could not execute CQB drills safely and properly discriminate or identify targets would be terminated. Zero tolerance. In Special Operations, we could not fail a mission by shooting a fellow soldier or an innocent hostage. It was simply unacceptable.

I remember going through training when the team was focusing on one of the members who was having a problem in CQB, and we would hesitate for a moment, watching him and his muzzle out of self-preservation before we would engage threat targets. It was unsettling for a week or so, but he got on board with the program and finished the training. In this case, had the instructors deemed it necessary, they could have removed him from training. The individual in question was probably being formally counseled without our knowing and had been walking a tightrope the entire time.

## ADMINISTRATIVE REQUIREMENTS

One of the biggest complaints I hear from the law enforcement community is that an individual belongs to a union or a civil service and they can't be terminated. This is a sad state that requires leadership to routinely document deficiencies and build a "packet" on the individual who is a hazard or safety problem. If you think the paperwork required is tough now, wait until you have a training accident or a problem on a raid and the investigation and law suits start. I strongly suggest that as a leader or administrator, you handle safety and discrimination problems quickly and decisively before they turn ugly and disrupt morale.

Resolution of safety problems or similar issues can take the form of formal counseling, or even reassignment. The biggest problem I have witnessed is the lack of counseling and paperwork when it comes to getting rid of someone not suited for the job. Generally, no one wants to hurt someone's feelings or put a black mark on their record, so the leader doesn't write anything down. He quietly lets this person slide on to another job, and things die down and people forget. Leadership will eventually change, and this person will continue to compete for promotion to be a candidate for a leadership position.

If an individual is in an assignment because he screwed up, resulting in a training accident, and there is no paperwork to back it up, someone did not have the nerve or professionalism to write it down. Now this individual with the problems is promoted and in charge of more people. I have witnessed this happen too many times. I prefer to call these people "leftovers," individuals that make it through the system because they have outlasted the leadership rotations and most people's memories. Sadly enough in this world, you must have it documented, or it does not mean squat. You can put someone away in a dark corner to work, and they will not come back to bother you, but they will come back to haunt someone else. So don't harm your current or future team mates by failing to lead, counsel, or take decisive action.

## TOTALITY OF AN INCIDENT

When writing a negative counseling statement, you should look at the following areas before you decide how hard to drop the hammer:

- Severity of incident.
- Worst-case potential outcome.
- Was it a training deficiency?
- Was it a failure to prepare?
- Was it a lack of attention, and why?
- Attitude and demeanor of person in question before and after the incident.
- Was this an ongoing problem?

Viewing the total information available and even recreating the incident in question is not too difficult to do to reach a confident conclusion to the questions that arise over the incident. Especially in difficult or questionable investigations, it is important that you sort through all the facts to ensure your picture is accurate. The question a leader must ask them is, "What is this person's career or life worth?" Beyond the paperwork, the leader needs to live with the decision or they will dwell on it for some time.

## TECHNIQUES AND COURSES OF ACTION

I have been fortunate to see a wide variety of counseling techniques and corrective action. The spectrum runs from verbal or monetary compensation to removal from the organization. Much more severe techniques such as imprisonment can be seen in the regular army.

What upsets me the most are people who waste their time and my time. As an ROTC instructor, I would bust my rear to help a cadet that wanted to help themselves. I would stay late or come in early or spend personal time if needed to get them on the right track. The ones I had no time for were the ones who would not help themselves. As they started their junior year, I would generally run three or four off out of the program the first month. I could generally tell who was serious and not serious and gave them a chance to prove it. If they did not take the opportunity, I pushed them out of the program. Why? Because it was more efficient to do it now rather than to deal with them for months, writing counseling statements and babysitting them, knowing full well that I would have to fire them down the road. This way, I could concentrate on the cadets who wanted to be there and give them my full and undivided attention.

I even had kids that pulled out late in their third year, claiming that their father was a full-bird colonel and the only reason they stayed this long was because of him. I was happy that they found their inner strength, grew up, and started making their own choices in life. They will be much happier human beings for it and the service will be much better off. The commander supported me on my appraisals 90 percent of the time, and that was fine with me. The more cadets I was allowed to cull earlier in the program, the less time I would have to spend later counseling and removing them. During my time as a trainer, I kept detailed counseling statements and detailed notes on my screwups to save me time, work, and aggravation in the long run.

As a trainer I also made a point to be aware of the influence which cadets can have on one another. A positive attitude will boost morale, while a negative attitude can spread like a virus. When personal problems arose with a cadet, I addressed those issues on an individual level first to ensure that cadet did not "poison the well" by negatively influencing the training of the other students (for more information, see chapter 7).

## KEY POINTS

- Counsel effectively.
- Counsel regularly, accounting for both good and bad incidents.
- If someone fails to make the grade after multiple counseling sessions, document and fire them. Do not pass that person onto others and do not continue to poison the well.

# 10

# REALITIES OF COMBAT AND TACTICAL TIPS

*War was ugly and evil, for sure, but it was still the way things got done on most of the planet. Civilized states had nonviolent ways of resolving disputes, but that depend on the willingness on everyone involved to back down. Here in the raw Third World, people hadn't learned to back down at least not until after a lot of blood flowed. Victory was for those willing to fight and die. Intellectuals could theorize until they sucked their thumbs right off their hands, but in the real world, power still flowed from the barrel of a gun. If you wanted the starving masses in Somalia to eat, then you had to out muscle men like this Aidid, for whom starvation worked. You could send in your bleeding-heart do-gooders, you could hold hands and pray and sing hootenanny songs and invoke the great gods CNN and BBC, but the only way to finally open the roads to the big-eyed babies was to show up with more guns. And in the real world, nobody had more or better guns than America. If the good-hearted ideals of humankind were to prevail, then they needed men who could make it happen.*

—Mark Bowden

- **AGGRESSIVENESS—IT WILL KEEP YOU ALIVE**
- **SURGICAL VERSUS COMBAT**
- **INTENSITY AND REALISTIC TRAINING**
- **COMBAT MIND-SET**

---

## SCENARIO: COMMAND POST

I went back out into the courtyard and found that the AC and several other leaders had come in off the street. My guys were still out on the street hooking and jabbing with the locals. The commander had a great deal of pressure on himself at this time and I was looking to him for some guidance, but he had too much information coming in. So much so that he could not process it all and give directions at the same time. Present were two section sergeant majors, another captain, and Pete, a TL who had come in and backed me up when I started clearing the house.

The volume of fire was picking up and it seemed that we were receiving much more than we were giving. One of the section sergeant majors told me to go check on my guys and it pissed me off. I knew my guys were okay; my ATL was with them and I had left them for an hour before on a different hit with another TL. It was obvious that the additional leadership had nothing positive to input but would rather sit around the AC waiting for any scrap of information he would put out.

We had one wounded soldier from our security force lying on the floor of the CP and guys were starting to treat him. I was pissed, so I grabbed his rifle and three of his magazines and an M67 fragmentation grenade from one of his ammo pouches and headed outside. The same section sergeant majors asked me in an excited tone, "You're throwing grenades?" I told him I wasn't getting paid to bring them back.

I turned around and I made an immediate left-hand turn and moved down the wall about fifteen feet. I stacked the magazines in the dirt next to my right knee and told my guys to get in the CP. They moved past me and I started to do some shooting. I used the wounded soldier's rifle because it would shoot as well as mine and I did not know how long we would be there. Also, I did not want to get my rifle dirty until I really had to.

The M16A2 had a good feel to it with its long sight radius, so I started to clean the street to the north. The road to my front dipped down to the next intersection and then began to rise to the next cross street. I had a large leafy tree thirty-five yards to my left front, which blocked the enemy's view of me from a great portion of the road to the north. I started with deliberate aimed fire and began to put bullets into any location that could hide a human being. I started by putting rounds into the leafy tree to my front just to make sure that no bad guys had climbed up into it and were waiting for the right time to strike. I remember reading accounts of Japanese soldiers doing this during WWII and I just wanted to keep the locals honest.

Below the tree to my front were four to five soldiers who belonged to our blocking force and were gathered around a car that was parked next to the tree. A tin wall surrounded a courtyard on the left side of the tree. I did not like how close that courtyard wall was to the guys in our security position.

I grabbed the M67 grenade, flipped off the safety catch, pulled the pin, and threw it over the building, deep into the courtyard to keep any potential threats at bay. I waited and nothing. The son of a bitch was a dud. I grabbed one of the foreign grenades that we used (it had the best fragmentation pattern of all grenades tested). I pulled the pin and gave it a heave in the same spot. A few seconds later it went off with a nice soft boom that a composition B explosive makes.

I started working over the street again, engaging near targets and then moving farther down the street. I saw a tin shed next to the tree and began to engage it by putting two to three rounds through it at chest level and then working the floor over. My logic was that the high rounds would either hit them or drive them to the floor and the floor is where they would get hit. I then turned my attention to the street and put one to two rounds into every dark doorway or window in my field of view. About two-thirds my way up the street, I observed a wood picket fence made of three- or four-inch small vertical round logs or large branches. I methodically worked it over and moved to my next area of interest, which was the intersection thirty yards past it.

"Darters" would dash from one corner to the next, running like their hair was on fire. I would only be able to get three rounds off during their sprint. I would fire one on the leading edge of their body, one about eight inches to their front, and another about one and one-half feet in front of them. I was hoping that they would run into one of the three bullets. I was not worried about killing them but I'd rather take a good chunk of meat out of their body and discourage them from coming any closer to the fight. It was about a two-city-block shot, probably 150–200 meters, and I could never tell if I hit anyone. At that range, they would make it across the street and then probably bleed out among friends. I would then start my shooting sequence of windows and doors and move back to the closest targets.

I remember looking to my right at the pathetic-looking Black Hawk helicopter, lying on its side in the alley. There were about four or five security personnel around it, doing their best to pull security amid the oncoming fire from the alley to my left. They had even resorted to pulling the Kevlar floor plates out of the bird and propping them up as shields in front of them in the dirt street. I thought to myself, *That is crazy*. I watched as one security man, on a knee, took a dead-center hit on his chicken plate and got knocked backward. He was tough; he got back up in the same spot, and guess what? He got hit again in the plate. He fell backward this time and Darwin's theory kicked in and he scrambled for cover. He was a lucky man to take two hits on the chicken plate. If only one round would have glanced off his plate, it would have screamed through the Kevlar vest and ruined his day. (You see, the plate is the only thing that will stop rifle fire. The vest itself is good for only pistol rounds and fragmentation.)

During this time, I was firing about a foot and a half over the security position to my front that was huddled around the car. They appeared to be weapons tight (meaning that they were not shooting). I was pissed and was cussing, screaming, and shooting, all at the same time. After I fired my third magazine, I decided to go back into the CP and get some more ammo. I reentered the CP, and as I did, some bad guy was trying to put rounds on me. I went back to the soldier on the ground who had so graciously allowed me to use his rifle and took three more of his magazines off his gear and headed for the gate.

Rick said that I should be careful, that someone was trying to hit me, and I made a fast and hard left to my sweet spot next to the building where I was before. I began my same pattern of shooting and I remember a woman running with an apron held out straight and it appeared that it was weighted and contained something important. She was going from the close intersection left side, running to the right side of the street.

Well, the ROE was now a little more lenient. She was coming to a gunfight, probably bringing ammo, and that was wrong. Several positions started to light her up and I screwed up and started tracking her head instead of her body and I broke the shot. I think it was a tracer, and it went about an inch or so behind her head. One of the guys at the blocking position turned around and looked at me like I was crazy, and I screamed, "What the fuck are you looking at?"

There is no way he could have heard me because of the fire and all the stuff that was going on. He could probably tell from the look on my face that I was not in a pleasant mood.

Looking to my right, I remember seeing a soldier in the middle of the street on a knee about five feet from a doorway. It looked like someone told him to go there and it was a stupid idea and poor position. I had a tree for concealment, but he had nothing but two city blocks of positioned enemies looking at him. I screamed and motioned for him to get out of the street and back to the doorway, but he either ignored me or was just following orders. An enemy gunner zeroed in on him and the rounds started impacting around him in the dirt until one connected. He held fast until hit and then rolled over. Shit.

I glanced up and saw the guys from the CP were ready to dash out and get him. I started to lay down suppressive fire for their move, and as my fire picked up, they knew it was time to go. Two to three guys launched fast, grabbed him, and dragged him back to the CP. They began treatment and then called for an air force PJ (pararescue jumper), who was working on others across the street at the downed bird. I saw he was going to make a dash from there to our position, so I started working over the street to the north, giving him some covering fire. He was moving as fast as his feet would carry him with all his gear. He made it safely and started to look over the new patients. Looking right, I saw another soldier come out into the street, to about the same spot where the first was hit. I thought to myself, *Not again*. I screamed and motioned, but he stayed and the entire episode repeated itself. This time, the guys from the other side of the street dragged him back into their building.

I glanced to my left and saw something fly over the tin fence from the courtyard I had thrown my grenades into. It left a small smoke trail and hit one of the security guys in the back, who was on the other side of the car to my front. Two of the men realized it was a grenade and scurried around the car to use it as cover. As luck would have it, it was a dud and did not go off. I looked at the tin fence a few feet to their left and started working it over with rifle fire. I would punch evenly spaced holes chest high across it and then work it over with a few rounds at ground level. It was starting to worry me that we were not putting enough fire out and the bad guys were getting ballsy and moving to within hand grenade range. It was starting to suck.

To add to my already-fine day, the rifle that I was using jammed up at about magazine number 5 to 6 and started to double-feed. This pissed me off and I moved back into the CP. As I did, the enemy gunner who was tracking me could see me as I headed in, dumped a couple of rounds off to my left into the dirt, trying to keep me honest. I told Rick as I came in, "I think you're right, he is trying to hit me."

This time I grabbed another fragmentation grenade and a light antitank weapon (LAW) and headed back outside. I was able to get safely into my position for a third time. I saw one of the guys from the position to my front looking down the alley to my left. I threw the LAW underhanded end over end in the air and it landed on his left arm and he jumped. I threw the frag to them also and said, "Shoot the motherfucker and throw the fucking grenade!"

It was time to start getting my issued rifle dirty as I had cleared the jammed rifle and left it in the CP. I started to shoot and work over the street again when I saw a gunman with an AK poke his head and barrel around the corner of the right corner of the intersection to our right front. The individuals in the blocking position to the front by the car fired a few rounds, but most impacted the side of the building and glanced off. I had two more grenades left and it was the perfect time to use one. I pulled the pin and launched it over the security position and just past the corner where the gunman poked out into the center of the street. It went off and I am sure it cleaned out anyone hiding around the corner. No one else poked out of that position again.

By this time, the air force PJ was running low on medical supplies and needed to make a dash to the crashed bird and bring some more supplies back. He was the man. As I started to shoot, he dashed back to the bird, scrounged around for a few seconds, filled his hands with what he needed, and then ran back again at breakneck speed. He did a helluva of a job. I pulled off my position and reentered the CP and held by the front gate, using it as cover.

I started to think my luck was beginning to run out hiding behind the visual barrier of the tree, and at any time someone might decide to shoot through the tree and get a piece of me. I tucked back in the gate and started to send a few rounds downrange using the doorway as cover. My buddy started to help me out by shooting a foot over my head and about a foot back. His muzzle blast about snapped my neck, and I thought I was hit with an RPG. I turned around and he was laughing. I told him to get his ass over the top of me if he wanted to shoot.

Wanting to see what progress the force was making, I left Rick at the gate and moved to the AC. He was in information overload as his "command cell" was gathered around, doing nothing but stealing oxygen. I started to look around in the courtyard and found something that I did not like. There was a building attached to the courtyard, a two-story that overlooked the courtyard that had not been cleared. It would have been easy for the bad guys to get in there and simply drop grenades into the CP. Life would suck if that happened.

Seeing this, I tried to give the AC three simple points to improve our position. The first was to tie in all the elements before it got dark and to establish fields of interlocking fire. Next was to clear the two-story building and make sure the bad guys were not there or to let them occupy it. Finally, I wanted to take my team across the alley to the building by the car/tree and establish a security position there.

I wanted to be able to push our perimeter out farther and make the bad guys cross a great deal of open area to get to us. This would make it easier for us to see them, engage them, or direct fire support on them. He said, "No," apparently I needed to go across the street toward the crashed bird and link up with the CSAR element instead. I told him that it would be better for me to push out the other way because of the fields of fire. We went back and forth a couple of times and I said, "Fuck it, you're in charge."

I gathered my team and held by the front door. I told them the plan and we could see two doors directly across the street from us. The left was open and occupied by friendly forces. The right was closed, and this was the one I was worried about. I did not know if any bad guys were trapped in there, and I wanted to clear it first. I held the team inside the gate until the sun started to set and it got dark. Once it got dark enough so that it was difficult to see the next intersection, I knew the bad guys would have a hard time seeing us. I gave the signal and waited for the squeeze. I led out and hit the door, clearing the small room first and then a second room. A barred window connected our team to the team I had seen in left doorway. It looked like this would be home for the night.

## AFTER-ACTION COMMENTS

### *SUSTAIN*

- Keep up aggressive action and momentum.
- Use accurate semiautomatic fire to suppress the enemy.

### *IMPROVE*

- Ensure soldiers use cover at all times and leaders understand what cover is when putting soldiers into position.
- Rehearse hasty defense actions dry fire and live fire, to include call for fire missions.

## AGGRESSIVENESS—IT WILL KEEP YOU ALIVE

> Wade through the wounded and vault over the dead.

I took this old saying from a dear friend, Master Chief Hershel Davis, an "old frogman" as he puts it. It is fitting and sums up the mental and physical attitude that you must strive to attain when preparing for combat operations. It is not enough that a soldier go to war. They must go with a positive and aggressive mind-set that ensures when they are shot at, they're going to hit back hard, wade through the enemy, and look for more on the other side. Plan on doing this time and time again.

## SURGICAL VERSUS COMBAT

Everyone has a different concept of when to change from surgical to combat. The rookie police officer or new soldier on the ground will have a different perception from that of the veteran sergeant. If the sergeant or junior leader has seen their share of action, their tolerance for allowing their soldiers to get hurt will be less than an inexperienced leader's. I have no time for a leader who would put their soldiers at a disadvantage over ROE or who puts soldiers' lives in jeopardy to comply with political directives. I have seen this happen too many times in life and have seen the "gray" area where the U.S. State Department comes in and tries to dictate ROE for troops in the negotiating phase of an operation. You have to make the peace before you keep the peace, and the safest way to keep the peace is to control with a heavy hand until you leave.

Too often we want to release our grip and drop our security in a country and this is when we get hit. This routinely happens when operations go from Department of Defense (DOD) to Department of State (DOS). I never thought I would see Americans killed, dragged, and mutilated in the streets of a foreign nation, twice in ten years. First it was in Somalia and we made the mistake of doing nothing. Then it happened again in Iraq and our leaders were too weak to take action again. They would rather sacrifice a few than make "political or military waves."

Leaders should know which troops are able to perform surgical missions and which are "combat troops." Average troops without any depth of training will have a difficult time performing surgical missions and are more likely to incur more casualties when attempting such actions. They should be used for what they do best, combat actions where they can employ the force that needs to accomplish the job.

If you need surgical work performed, call a surgeon or Special Operations soldiers and let them do the job. They have the time, the training budgets, and are especially selected personnel to accomplish the mission with minimal casualties. Allowing them to do the job they rehearse and train for and prevent casualties.

## INTENSITY AND REALISTIC TRAINING

Once you master the basics, it is important to begin to reinforce the training with simple and realistic scenarios. The visualization phase of training needs to be confirmed with mission-oriented and realistic scenarios which require the individual to work with all the tools they carry on a mission. This training ensures that individual carries the proper gear and in the spot they can most efficiently use it. For more detailed exercises and tactical techniques, see part II of this book.

## COMBAT MIND-SET

It may seem repetitive, but I cannot overstate the importance of having the correct combat mind-set. There are no guarantees in combat. I watched a highly trained Special Operations soldier take a ricochet off a wall into his forehead from an untrained soldier carrying an unzeroed AK-47, several blocks away. All I can say is that it was his time to go. Occasionally, if the bad guys sling enough lead, you will get hit if you try to move through it. Be smart, go around it when you can, and stay focused. Prepare the best way you can, and give it your best shot. Trust your instincts. There are no guarantees in life, but train in a system that will give you the best chances of survival and then allow your controlled and aggressive combat mind-set to do the rest. Once you come back, honestly evaluate what and how you did and implement changes to make it better.

## KEY POINTS

- Be aggressive.
- Always stay in the fight and never give up.
- Always look to better your position.

# PART II

## TACTICAL INSTRUCTION AND LEADERSHIP IN THE FIELD

# 11

# DEVELOPING YOURSELF INTO A GOOD STUDENT

*An open mind, humility, and enthusiasm are a good start.*

- **DEFINING OBJECTIVES AND SETTING GOALS**
- **SELF-DISCIPLINE**
- **YOUR PATH**
- **KEEPING AN OPEN MIND**
- **RECORDING YOUR JOURNEY**
- **DEVELOPING A ROUTINE**
- **SIMPLIFYING YOUR PERSONAL COMBAT SYSTEM**
- **CONDUCTING HONEST EVALUATIONS/ SELF-ASSESSMENTS**
- **ATTENDING TRAINING**
- **REMEMBERING WHY SOMETHING DOES NOT WORK**
- **CONCLUSION**

## DEFINING OBJECTIVES AND SETTING GOALS

As a new student aspiring to be an instructor, you must choose objectives and set goals in your quest to become an instructor. Do you want to know a great deal about a specific area or a little about everything? Or, do you want to have a broad knowledge about the entire workings of your profession? Think about an old-school carpenter who becomes a contractor and knows the ins and outs of their job versus a new contractor who must rely on "subs" to perform the work without the in-depth knowledge of their quality of work.

If you wish to specialize in one field such as being an urban marksman or sniper, you can do that. You can go to schools ranging from military to civilian or take private lessons from private instructors. All paths will get you there. You can be as good as you wish to be. I also believe there is room for "professional corporals" as in the British Army. These are individuals who love what they do and do not wish to attain rank and move away from their passion. Their instructional experience will be limited to a specific field, but there is no problem with this. Also, they can make good battle buddies for new team members.

---

### SCENARIO: A CARPENTER

My father started his contracting career as an apprentice carpenter. In the 1950s he began his apprenticeship as a cabinet maker. He worked as an apprentice for several years and learned the trade. In those days, the carpenter's crew built the house from the slab up. They knew about concrete, framing, roofs, etc. When their time came to transition to the job of general contractor, supervising the "subs" or subcontractors, they had a good grasp of what everyone's role was, what the standard was, and what constituted shoddy or poor work. Now, the key to building houses is to hire subs that are fire-and-forget. They specialize in their areas such as framing, roofing, cabinets, etc.

This analogy can be applied to your tactical career. You need to start out at the bottom level and learn everything there is to know about your job. I will use mine as an example. In my day, I started at the bottom of the team and my first job was team breacher. I had to learn all methods of entry and had the team's current breacher there to oversee my training. Even though I thought I knew a great deal from my introductory training, I had a long way to go to meet team standards. Eventually, I attended a three-week in-house breaching school that covered many areas that were missed during my basic phase. With every hit, I reinforced my training and saw something different. I catalogued in my mind what worked and what did not. I checked my equipment and rechecked it prior to training exercises and during day-to-day operations. I never knew when I was going to get called out and see something new.

Your job as a breacher was critical, even though you were the bottom guy on the team and most of the time last in the door. If you could not successfully breach, you could not get in to get the bad guys. You might think that you are the "bottom-feeder" of the team and you never get "trigger time," the reality is that you are extremely critical to the operation.

You need to put this in your mind and understand that your team and TL are relying on you. Your sister teams are relying on you and the entire force is relying on you to do your job and do it well. Knowing what to do if the charge succeeds is critical, but just as important is knowing what to do if it does not succeed. If the charge does not go, it is your job to render it safe. You must also be prepared with manual or mechanical breaching tools should the door hang up or become stuck. If the door does go and there is a wall of garbage behind it that you cannot get through, you must know your alternate breach point. You must know all your equipment and how the sun, rain, freezing cold, hot, and humid environments affect you and your tools, and you must prepare for these contingencies. How do you learn this? By training in every environment.

You need to practice all your breaching skills in jungle, arctic, desert, and wet environments to know the environmental factors that will affect you and your equipment. You need to know how to get up to your breach point so you can breach your door in the sun, rain, and snow, day or night.

In the end, put your ego aside and know whatever job you do is critical. Also know that one day you will be point, leading the tip of the spear, and your breacher will quickly and efficiently get you into your target.

---

## SELF-DISCIPLINE

Generally, your mind will race when you begin your dynamic career and you will want to know and do the "cool" (a.k.a. action guy) stuff to satisfy the adrenaline junkie in you. Realize that not all things are as easy as they look. Let us take learning CQB. Some people will look at a video clip and say, "That is easy, I can do that," not realizing that there are many subtasks or skills needed to get you to that level of performance.

First, you need to be in shape. If you do not possess some level of physical fitness, you cannot sustain yourself wearing your assault gear, for hours at a time, while learning to shoot. You will need to do countless runs both day and night to develop your skills. If you are not in shape, you will be thinking about how bad your body hurts and not what you need to do to make the next run better.

Learning the basics of shooting with a pistol and a rifle and all the safety protocols that go with these skills is the next order of business. You need to learn a discrimination system that teaches how to look at people and targets to ensure that you do not shoot the wrong one. The next step is learning how to move as an individual and as a group with weapons. Approaching and penetrating breach points are the next critical skills. You must also learn how to enter a room by learning to "read" the shooter in front of you, to know which way to go, and to "process" the room. That is, making split-second decisions and surgically engaging threat targets with other officers shooting near you. This takes weeks, if not months, to teach and learn and to elevate a student to a basic level of performance. Only through supervised repetition will you progress to smooth and efficient levels of performance. Each step in this process can take an initial day of training and several sessions of sustainment or maintenance training.

## YOUR PATH

You have to make a choice on which path to pursue. Once you make your mind up, develop a plan and go. Training is expensive. Military or police training proves to be the most cost-effective way to become trained in certain fields. Once in the service or police work, the service or department will send you to other training schools to further your development.

Regarding the tactical training field, combat time or street time should be a requirement to becoming an instructor. You should plan this into your path and use it to build your experience base for future instructor reference. Too many times, officers go through high-risk situations and say, "That was a rush," and do not take the time to reflect or record the lessons learned. As you are an instructor, students will seek you out to learn about your combat experience and how you applied this to your teaching and training.

Special units in the military, and the size of the departments in civilian life, will determine how fast and how much training and experience you receive. Working with a major police department will get you exposed to more high-risk situations on a repetitive basis in two years than a small department may give you in five to ten. This also may include training. Both small and large agencies serve the public good and have noble causes. The exposure rate to high-risk situations may be higher in a larger, busier department, adding to your tactical and experience database. It may cause an officer to mature and season a bit faster than in a smaller agency with less of a workload. This experience rate may help complement your instructor knowledge base (as a general rule).

There are always exceptions to the rule. I work with a great deal of small and midsized departments that have exceptional training programs and are on the cutting edge of training and information. In the end, research the agency or organization you wish to work for.

If you had a sheltered upbringing in a small town, as a new officer, you may want to start on a small or midsized department before jumping in the fire of a major metropolitan agency. The learning curve may be too steep for you mentally, physically, and psychologically. I cannot tell you where to start, but use your common sense. Talk to individuals doing the job and have them give you an honest assessment of your capabilities.

## KEEPING AN OPEN MIND

During your journey, you will be exposed to countless training systems, methodologies, techniques, and tactics. All have their place. Some are better than others. Be cautious on tactics or techniques that have no basis or solid background or use. Most reputable instructors will not teach a tactic or technique unless they have used it in either competition, combat, or the street. If your tactic or technique has not been validated in one of these situations, be cautious, if not suspect. Some of the best instructors have subtle differences in teachings and techniques. If you or your agency is paying the bills, ask questions and make them earn their pay. If you do not understand something, ask. If the material contradicts the current doctrine that you are employing, ask the instructor as to the advantages of their system. Most instructors will not be offended. I prefer students ask questions. It keeps me sharp and ensures that I know the material and retain the reasoning for its use.

There are few "smoke and mirrors" instructors who want to entertain rather than teach you. There is a bit of entertainment needed in the classroom to keep the mind relaxed, but if the class is structured on 80 percent war stories and jokes, it is an indicator to me that the instructor does not have the adequate background or information to deliver a satisfactory class. Should you get stuck in one of those classes, search the information for the one jewel that you can take back with you. It is said that "a blind pig can root up an acorn from time to time." Be that blind pig. Ignore the fluff, the ego, the stories, but pick up the one piece you can add to your puzzle and future goal of becoming an instructor.

## RECORDING YOUR JOURNEY

Computers and I crossed paths during the middle of my army career. I have been a slave to them ever since. I have learned to use them in a positive way to both organize and teach. I use them to record after-action comments and reports of classes of what did work and what I needed to fix. I like to write down personal training systems that I have used, how they worked, and where they got me. I may use them again ten years down the road in my life if I am repeating something like gearing up for a major shooting competition. The computer will preserve this data, save training time, keep me efficient, and keep me from starting from scratch on my next training endeavor.

Photos are another great way of recording your journey and saving time and effort. During my time in special operations, we did not have the operation tempo that today's soldiers have. We might train and rehearse for a mission one year and not see it for a year or two. Part of the problem with that would be the amount of time it takes to prepare your gear for rehearsals and the mission. Taking pictures of your gear, whether static on the ground or recorded as a simple team picture with the team wearing their equipment, will save you a great deal of time. You can quickly scan a picture of how you wore your breaching equipment or, if you were a shotgun breacher, how you mounted the weapon and where you put your spare ammunition. The same applies if you were a sniper, SAW gunner, or M203 gunner. It simply saves time and keeps you from reinventing the wheel.

As for taking and recording notes, I have used everything from business cards to voice recorders to take notes on points that I wish to recall from a training day or class. The hard part is to have the discipline at the end of the day to capture them and put them in a logical sequence and store them in a place for future use. Again, your computer can help. If you do not take the time to do it each day, you will probably find the index cards in a ball after the washing machine gets through with them and you will stare at the mess trying to figure out what important points you were trying to capture.

## DEVELOPING A ROUTINE

As I have gotten older and find myself spending a great deal of training time on the road, I have relearned the value of a daily routine that includes physical training, technical or tactical training, and mental training or stimulation.

Your mental state of mind—whether stressed, relaxed, focused, or sharp—is up to you and is probably the most important factor in developing and sustaining a routine. Without a disciplined, focused, and relaxed mind, you will not be able to develop and adhere to an effective personal routine.

Many self-improvement books discuss the importance of time and time management. As pointed out by so many, there is only so much time in a day and either you manage it or it will manage you. First, outline all the goals you want to ritually perform in the average day. They may include, but are not limited to, the following:

- Physical training
- Dry fire (rifle/pistol)
- Knife work
- Hand combat
- Professional reading
- Mental time (meditation)

The above does not include work and family time. Take the different areas and prioritize them in accordance with what is important to you. A married person may want to put more time and effort toward their family and kids whereas a single person may channel more time toward other areas.

Physical training can be accomplished in many different ways and systems, but probably schedule at least thirty minutes on the light side and one and a half hours on the heavy side for one session. Physical training was sometimes done back-to-back due to time constraints and my schedule, working both cardio and strength in one session. During my time in Special Operations, I would come in early and run before breakfast. I did my weight workout before lunch. You will have to figure out what schedule works for you.

Dry-fire, knife, and hand combat drills can be accomplished in a twenty- to thirty-minute total module. Dry fire can be done in a confined space with reduced targets on a bullet-safe wall. A "Bad Bob" dummy can be used for your knife and hand combat practice. You can begin your day or end your day with these drills. If I was military, in a hostile environment, or a patrol officer getting ready for the street, I would begin my day with these drills as I would be mentally and physically preparing for possible combat.

What should you practice? As you start your tactical training, you should be developing a common task, or *hard skills* list of basic drills, that you would use in combat situations. Over the years, I have collected rifle, pistol, hand combat, knife, tactical movement drills, etc. that I routinely practice or refer to and share with students. Unless you write these down, you will not remember and ingrain them. I dry- or live-practice my ten pistol standards, many times using a shooting timer with a second beep to ensure that I am meeting the standard. I can do the same for the rifle. It is up to you and your imagination. I knew of one patrol shift commander who made the entire shift unload their guns, hang their soft body armor on a hanger against their wall locker, and dry-fire on it prior to going on shift. This is a great routine. It prepares the mind and body for combat.

## SIMPLIFYING YOUR PERSONAL COMBAT SYSTEM

Few combat systems complement each other. You can spend years learning one stance in martial arts, and then learn a new one for pistol, rifle, etc. Try and find one stance that will work for any weapons system. This way, your platform will be the same for all systems and you will learn and establish it rapidly. I was able to learn from a variety of instructors who were very talented shooters. Some knew a very narrow bandwidth of information. For example, a world-class pistol instructor may have a system that he is very comfortable with and it works for him. The problem that I encountered is that the pistol is a secondary weapon and usually would not come out unless my rifle failed. Issues I wrestled with were the position of my feet and balance. I did not want to change my stance dramatically from rifle to pistol as this would cost me training time, efficiency, and muscle memory.

It took me years as an instructor to develop one system that complements everything in stance, balance, hand movement, etc. I had to reverse engineer my way of doing things and determine what would fit and what would not. If it did not fit or complement one system, did I need to keep it? Or, could I just let it go?

Stance, or my platform, was one of these points. It is critical to all that you do. I shoot a modified isosceles stance. My firing foot is back at a position where the toe of my firing foot is in line with the heel of my nonfiring foot. My feet are shoulder width apart. Everyone's shoulders are a different width apart, so everyone's feet will be slightly different. Weight distribution is going to be different for each individual. Everyone's body is structured differently. I lean slightly forward 70 percent or so on the balls of my feet and 30 percent on my heels. My shoulders are over the tips of my toes. My head is slightly forward of my shoulders. My upper body is forward of my lower body to help absorb recoil during multiple shot drills with rifle, pistol, or shotgun. I use a video camera and multiple-shot drills to help students learn their balance and weight distribution. By taping from the side during multiple shot drills, I can see how the body is reacting to being pushed rearward and how the students' posture/stance is either positive, negative, or neutral. Negative, or neutral means being pushed back and thus taking longer to realign sights for follow up shots.

Once I get the shooting platform down, I add a hand combat system that will work without moving the feet, possibly only shifting weight. I have found one such instructor that will dovetail his hand combat system into my shooting system and they will complement each other.

Dry firing or shooting your rifle will build or reinforce your pistol stance along with your hand combat stance. You might have minor weight transfers, but as I said, they will be minor. Core balance is core balance. In the end, if you can find one system that complements everything, you will progress faster in learning your tactical skills. Each time you assume your stance, you are practicing a stance that can be used for rifle, pistol, shotgun, hand combat, knife work, etc. You are simplifying and working one system that will last you the rest of your life.

## CONDUCTING HONEST EVALUATIONS/ SELF-ASSESSMENTS

Occasionally, test yourself cold and warmed up. Cold is when you go to the range, pull your weapon out, and shoot the standards with no practice or dry fire. This will show where you are without practice or notice should you have to react to a life-threatening situation. Do the same warmed up and see where you stand. If you want to take it a step further, video the evaluation or exercise and review it. Find two or three major problems and fix them before your next evaluation.

Use standards that are proven and goal oriented. If you prefer, compete in shooting matches to help induce stress. Just understand the difference between games and real-life training. The same applies to hand combat. There is nothing like getting hit to help you understand what it feels like. You also need to learn to control your emotions, fear, apprehension, etc. Understanding your physical strength and endurance limits is also important. Remember to train or practice with professionals in controlled environments and understand that this is not combat or the street. Videoing the session will help you understand what you are really doing and not what you perceive yourself to be doing. You will see the good, the bad, and the ugly. How and what you take from it is up to you. If you can emotionally detach yourself from it and say I need to work on this, this, and this, you will gain more from the video.

## ATTENDING TRAINING

This section could also be titled "How to Be a Good Student." While I was teaching ROTC in my early forties, I went back to school for my graduate degree. I was able to see both sides of the teaching equation as a student and as an instructor. Most of my graduate classes were held at night after a full day that sometimes started at 5:00 a.m. Needless to say, I was tired when I went to class.

I did simple things to make life and learning easier. I purchased a separate notebook for each class to keep the material organized.

I put the syllabus to the course up front and zeroed in on the grading curve and established what was important. I kept a master calendar of all projects due and started a backward planning process as taught in the military. I planned when I needed to begin each project and ensured that I had enough time to get a rough draft to each professor to proofread prior to submitting a final paper.

Too many times students submit papers that are off the mark and do not address the topics or answer the questions the professors want answered. All professors I worked with were more than willing to scrub a paper in advance. They were happy to have a student that was proactive rather than reactive, full of excuses, or did not care. This technique ensured I received the desired grade. When I studied for classes, I went to the university library. I could not concentrate on the material in my home with the day-to-day family activities going on. There were just too many distractions. The school library is the most underused building on the college campus. At our school, it only got busy one to two weeks before finals. The rest of the time it has numerous quiet areas where you can get twice as much accomplished for the same amount of time spent at home.

### *Your Body*

As for a tactical training classes, come ready to train. First, be in good, if not great, physical shape, so that any routine demands of training does not affect learning. Bring all the inclement-weather clothing and gear you need. If you are thinking about how cold it is, you are not focused on training. If you are dying of thirst, you have not properly hydrated. Bring water, snacks, and food to keep your body going. I have even gone to the lengths of preparing medical kits with basic prescription drugs in case I get sick from allergies or headaches or colds.

### *Your Mind*

Did you bring a notebook and methods to take notes? Pen, pencil, or voice recorder? In short, take notes. I hand out notebooks in every class with a note-taking section. At the end of the week, most are still blank. Take notes and review notes at night to ensure you understand the material. Few students take notes or highlight critical points. I find that students who are serious take notes. These same students are, generally, the ones paying their own way to the training.

Review the material at night and bring in any questions you have to the instructor the next morning. I always start a class by asking if anyone has any questions on the previous day's material. Rarely do I have any questions.

### *Your Equipment*

Is your equipment clean, serviceable, and present? Have you packed it all? Are your guns zeroed? Do you have ammo? Do you have cleaning gear? Do you have batteries for electronic devices? As an instructor, I try and bring extra for those students who forget. When I attend firearms training, I also carry spare guns should mine break during a course of fire. All I have to do is pull out my spare gun and keep shooting. I can fix the broken one after class. This way, valuable training time is not lost. Also, take the time on breaks to adjust or modify equipment that helps you or serves a purpose. Add equipment that you need and discard equipment that hinders or restricts your movement.

## REMEMBERING WHY SOMETHING DOES NOT WORK

Exposing yourself to many different schools of training may frustrate and confuse you at one point in your career. You may have to catalog the techniques for outside movement, inside movement, CQB, etc. Some techniques may work for one venue and not another. Streamline the techniques you use and practice into a simple and comprehensive package. If not, you will have a cluttered toolbox that will not work for you. You will not find the right tool when you need it.

With the various techniques you learn, note why they are not the best fit in your toolbox. You will see them again one day as an instructor. Students and instructors will reinvent the wheel by bringing back old discarded techniques, as new or new and improved. As an instructor, you must remember why you choose not to use a certain technique even if it looked simple and efficient. It may also have some safety issues that build bad habits.

One that comes to mind is that of keeping the finger on the trigger while transitioning between multiple targets on a flat range. Yes, speed shooters will say you have to do it to save time. In reality, as a law enforcement or service member, you may have to sweep across innocents or even friendly officers during a gun battle. It is much safer for everyone if your finger is off the trigger should someone shoot close to you or a flash bang or a breaching charge go off. You might accidentally squeeze the trigger and discharge a round when you did not want to.

As a final story, I attended a prominent shooting school for the second time in my life as a student. The instructor was a great guy, but did not have a tactical background and had not been to combat. He was teaching a shooting and moving drill and I was a bit slow in learning it. He suggested that I prep my trigger prior to making the shot as a technique to get the trigger pull down. I kept my mouth shut, but all sorts of warning lights went off in my head. To execute this drill properly in a tactical situation, you would have to move with your finger on the trigger and have the trigger partially engaged. This was unacceptable for the training that I taught. I practiced the drill at my slower pace and kept the mental note to this day. I knew I could never teach officers to move down a hallway or to a breach point with a finger on the trigger and the gun off safe. It would be criminal to do so. For all practical purposes, the drill was useless.

## CONCLUSION

Take the time now and press into your mind that the route to being an instructor is a lifelong journey. You will continue to be a student and keep adding to your knowledge. With that knowledge, you will continue to refine what you teach and to put it into clear, concise, and logical packages for your students to absorb and digest.

## KEY POINTS

- Do you want to specialize or understand the entire field of work? Set your goals and path accordingly.
- Record your journey.
- Be the blind pig.
- Develop a routing and adjust as necessary.
- Develop a simple system that will complement itself.

# 12

# DEVELOPING A TEACHING MIND-SET

*Teaching is the greatest gift you can give back in life. Helping others attain and learn what you have learned, more efficiently, only with less drama.*

- **WHY DO YOU WANT TO TEACH?**
- **LEADERSHIP SHOULD BE A REQUIRED CLASS**
- **FOCUS ON WHAT IS IMPORTANT: SAFETY, TECHNICAL SKILLS, AND TACTICAL TRAINING**
- **BE HUMBLE**
- **BE THICK-SKINNED**
- **LEARN TO SAY I DON'T KNOW**
- **LEARN HOW TO PUT STUDENTS AT EASE**
- **RECORD YOUR SUCCESSES AND WEAKNESSES**
- **KEEP AN OPEN MIND**
- **MAKE THE NEXT CLASS BETTER THAN YOUR LAST**
- **THE HARD CASES**
- **GOLDEN RULES OF TACTICAL INSTRUCTION**

## SCENARIO: THE "Q" COURSE

I was fortunate to serve under a man in the late 1990s who I thought was a great leader. When he was in his midforties he went through the Special Forces Qualification Course as a student. The extraordinary point about this was that he was a full-bird colonel who was taking over command of the entire Special Forces Training Division and had already completed the training. When asked, "Why are you going through as a private?" His answer was simple, "If I don't know what is broke, I can't fix it."

I remember this same officer as our section commander, always leading by example. One day we had a back-to-back gut check that consisted of multiple events. We started with a multievent round-robin in the gym and then moved outside for more physical tests. It was hot outside, 105–110 degrees and we grabbed our ruck-sacks and did an eight-mile road march. It was so hot that MPs stopped and asked us if we had permission to march in the heat. We ignored them and kept on trucking. The march ended at the obstacle course where we went through the twenty-six obstacles and then moved back to the main building. There we reported to the pool and swam five hundred meters nonstop to finish out the morning. This mini–gut check was to be one of many that I fondly recall. Since it was geared toward the individual, no one yelled or screamed at you, you could move at whatever pace you deemed appropriate which would give you enough energy to complete the next phase. You always had to be smart about this. If you put too much energy into one event, you might fail on the next one or one more down the line. You never knew how many more events were to come, and this helped strengthen your mind for the unknown, and it prepared you to deal with the unexpected. You learned to move at about 80 percent speed, always keeping a bit in reserve. Being tired actually helps your wandering mind by taking away all the excess energy that causes you to lose focus. It also helps you to take one step at a time and focus on each step.

AFTER-ACTION COMMENTS

*SUSTAIN*

- Don't be afraid to get your hands dirty.

*IMPROVE*

- Find more leaders as described above.

---

## WHY DO YOU WANT TO TEACH?

Before you can begin teaching, you must ask yourself the most basic question: *Why do you want to teach?* Are you driven by ego, your self-esteem? Do you merely like to hear yourself talk, or do you have a genuine desire to pass information along?

Take the time to first figure out why you want to teach. Is it to bolster your ego to feel important and have people sit back in awe of your stunning presence? Or, does it look to be an easy road and steady paycheck? Either of these reasons does nothing for the teaching profession, let alone your ego. First, students will easily see through your shallow veil of knowledge and soon understand who and what you are. A poser of sorts. A poser is one who claims to be someone or something they are not.

However, if you are looking to honestly give back to society and to help pave the way for more efficient learning, then take up the profession. If you are handed a lesson plan, make it better than when you received it. Then, hand the lesson plan and class to your replacement better than when you found it. It only requires a bit of work ethic and pride.

The primary job of a teacher is to funnel information for your students. The teacher should take all the acquired knowledge and condense it down into an easily digestible format that a student can easily learn. My job is to sort out what is important and relevant, and what is not. Let's face it; students have a great deal on their educational plates. In the tactical world, I don't say, "Know all the techniques, good and bad." I say, "Here are the techniques I believe in and why and this is why I do not use these other ones. These are the ones I want you to know and practice." Students are paying me to take my real world experiences and knowledge and tell them what works and what does not. I must do this in an easy and efficient manner, otherwise I am lazy. I failed to prepare and I cared only to do a mediocre job.

## LEADERSHIP SHOULD BE A REQUIRED CLASS

Are instructors leaders? You bet. They take control of a group for a determined period of time and move them from one goal to another and their influence and style can have a great impact on the momentum and strides that the class has. I have found as an instructor that when you are assigned to teach a subject and you take the time to properly prepare, you will learn a great deal more than the average instructor will who is simply out to punch a ticket. For example, when I teach instructor courses in the law enforcement community, I require students to prepare and deliver a block of instruction to a class of new students and to sit through all the blocks of instruction.

This technique does several things. First, it allows the instructor students to see the material several times and the class twice. The students see the material during the instructor class again when they study at night, and again during the student class when they either deliver the block or listen to another instructor who gives the block of instruction. Repetition is the key to grasping and understanding the material. I generally find that if an instructor puts out a great deal of information, the student generally only retains 50 to 70 percent if they are lucky, and, if they take notes. I see a great deal of students that don't take notes and probably only get up to 50 percent of the information if they are lucky.

I believe that leadership should be a required class and that students should be required at times to study it, regurgitate to an audience, and then put it into practical application. Leaders have the ability to impact work and life issues and to have a positive influence on both. The same goes for the impact they have on the lives that they touch and influence in a day-to-day work and leadership role. They have the ability to make the workplace a fun and enjoyable experience or a daily dull grind.

As you read over the tactical training methods I describe in this book, keep in mind the fact that you should use the same style and system in the administrative world as you would use in training or combat. Nothing should change from training to combat or administration to business. The staff that is designated to support you should have the ability to ensure that training and training exercises are set up and run at a very high level. Exercises should be routinely run to ensure that command elements are versed in the proper solutions for the tactical problems they will face. In the military, simple range fire exercises or complex targets will be used by good leaders, and they will use the available assets to ensure soldiers get the most out of every exercise. Civilian leaders can easily do the same in their corporate training.

## FOCUS ON WHAT IS IMPORTANT: SAFETY, TECHNICAL SKILLS, AND TACTICAL TRAINING

Leaders and trainers need to be technically or tactically proficient in their skills in order to establish confidence and earn the respect of students. If an instructor cannot perform the task, or at least walk through it with an air of confidence, students will not believe in them or their methods. During my Rifle and Pistol Instructor classes, I perform demos of some drills, not all. If a student requests a demo, I will give them one at a slow speed, and then perform the drill at a combat speed, so they know that it can be done. I also shoot the standards with every class so they see it can be done. I wear all of my gear, including helmet, goggles, and gloves, to prove it can be done in equipment and you can train to the same standard. During instruction, I also use a standard rifle and iron sights. It not only shows it can be done, but takes away any excuses the average student may wish to use as to why they cannot do the same.

Maintaining my proficiency keeps me humble and practicing the basics. It ensures that I keep with current training trends and techniques, so that I know what the students are going through in their mind. It helps me to look for different ways to more efficiently teach students what I know and how to do it.

Finally, I hate excuses. When a forty-nine-year-old man can put on all the gear and shoot the standards without any practice and qualify on the first or the second try, it takes all the potential whining out of students who are used to sniveling for a grade instead of working hard and earning it.

Just as importantly, leaders and trainers need to understand the technical, tactical, and safety aspects of the job, so they make more informed and intelligent decisions on a course of action. If a certain course of action seems risky, then the leader must have the knowledge base to approve, disapprove, or help modify the plan for successful implementation. Whether you're a police officer who has been called to respond to an incident or a soldier on assignment, you need to be aware of all your logistical assets, such as trucks, squad cars, department helicopters, etc. Should someone be hurt or killed, you need to have a medical support plan in place to provide extraction of your casualties. Ultimately, if you, or your students, do not understand the tactical aspects of a situation, your men could miss an important detail in the planning phase and become trapped or expose themselves to more danger than necessary.

Safety is also one of the most important factors in this equation. Even in combat, safety measures need to be adhered to. Friendly fire or "fratricide" can induce as many casualties as enemy fire. It happens in every conflict, and it takes proper planning and identification to ensure that friendly troops don't fire on each other in the heat of battle. During the invasion of Panama, we almost lost more soldiers from friendly fire than from enemy fire. Individuals may argue this point, but I have witnessed several friendly fire incidents firsthand that were covered up and reported as hostile fire to save face for the unit as well as the individual who pulled the trigger. As the lethality of our weapons systems grow, so must our training.

## BE HUMBLE

*Closing a student's mind should be a criminal offense.* If you let your ego get in the way of your teaching, you may become flustered or angry and lose sight of your goal to enlighten or give back. I use the example of the egotistical military instructor who said, "I hope you all fail," as an introduction for students to a high-stress military school. His arrogance and cheesiness soon wore off. Students then looked for the "real" instructors in the course who were there to teach. We quickly realized that this instructor was simply smoke and mirrors and lacked confidence in himself. I have found that students quickly see instructors for who they really are, either a professional trainer or a person who needs to stroke their own ego and tries to impress the class with war stories, past deeds, etc.

Use stories to illustrate points, both good and bad. Use 75 percent from incidents you have seen and maximum 25 percent from incidents that you have been involved in. Let students tell their war stories and have a chance at being a voice. It will serve you well. This attitude, coupled with the ability to perform the technical tasks, will work wonders for you and your students.

Remember, it is not your job to compete with your students or to get your students to compete with each other. First, students must learn to compete with themselves and figure out what they must do to be successful. If they are watching a fellow student, they are not concentrating on their weaknesses. If they are competing with you, they may be bypassing the fundamentals to cut time, and it may be unproductive in the end. Encourage them to "watch their own lane" and concentrate on their own weaknesses and shortcomings in an effort to become better.

## BE THICK-SKINNED

Students may come to class with attitudes that you do not appreciate. Understand that they may have not had your upbringing, life experiences, etc. Give them a chance to see the error of their ways before you nuke them, especially in public. If you attack them personally, you will probably have an enemy for life and have just closed their mind.

This comes into play in my business when dealing with tactics. Everybody knows tactics, and it seems he who knows the most wins. Most people do not know the history of the tactical drills they have learned. Sometimes, the techniques are so diluted over time, they are useless.

I remember working with a prominent tactical team when I first started. They looked at me hard from day one. When I started talking about tactical exterior movement, they were thinking to themselves, *Who is this guy*? This team was used to going from point A to B in a file or stack. I taught something radical, a heavy head (two guns/sets of eyes to the front), then a flow into an immediate attack formation when contact was made. Since we had some extra time, I allowed them to use their technique first. We filmed it and counted the number of hits on friendly officers. We then used my technique, captured it on video, and again, counted the hits. We reviewed both techniques in the class, via video projector, and received feedback from the role players and team members.

Both techniques would work, but by using my movement formation, less officers, were hit by bad-guy fire, the bad guys were hit by more officers, and the officers could react much faster and secure the area sooner than with their technique. In the end, I learned to use the phrase "Let's watch the video" when asked if a technique was good or bad. Students learned that while they performed the drill, their mind only saw parts of it. The video showed the big picture in a nonconfrontational manner, not making the instructor or the students the bad guy and keeping their minds open. In the end, they would use my technique of their own free will and believed in it. My success was found by taking the time to educate students in a nonconfrontational manner.

## LEARN TO SAY I DON'T KNOW

There are times in my life when I just had to admit that I did not know everything or just could not remember. As I get older, it seems to become more frequent. It goes along with leaving your ego at the door, but I do not have a problem saying, "I don't know the answer to your question, but I will research it and get back to you." I have learned that people appreciate the response and the admitted humbleness and are in awe when you get back to them with an answer or with a person to talk to who does know.

## LEARN HOW TO PUT STUDENTS AT EASE

There are many techniques to putting students at ease in their learning environment.

Remember, it is their world and you must fit into it. Some will be naturally relaxed. Some will be apprehensive. Some will be simply scared to death and overwhelmed.

It is said that confidence is contagious. I also believe that composure and panic follow this same rule. Breed confidence into them through your teaching and interaction. Show them how to train, study, rehearse, and then put the ball into their court. Let them know up front that they control the outcome of their action and inaction. Teach them to be strong willed or strong-minded, in the right cases, and not be stubborn to a fault.

Use a consistent professional teaching style. Use humor when it is appropriate. Use other students to help teach. Sometimes, students relate to other students as they do not see them as a threat or adversary.

## RECORD YOUR SUCCESSES AND WEAKNESSES

There have been a few times in my life where I had a great thought and I did not record it. Also, I sometimes found a great teaching methodology and did not record how or why I did it.

I kick myself now, especially while writing this book, for not having those notes. Record your strengths and positive experiences. Dwell upon them to the point where you can build on them and continually put them into practice in future lessons. Note weaknesses, then, try not to dwell on them and do not put them into action again. Learning not to do something again is learning.

## KEEP AN OPEN MIND

Keep an open mind and listen to students. If you get in the habit of cutting them off and not listening, they will take it as rudeness, rejection, or you are just a jerk. Either way, you are on your way to closing their minds. By taking the time to listen, you can formulate your response in a clear, articulate, and nonthreatening manner that will keep the students' minds open along with the clear lines of communications.

## MAKE THE NEXT CLASS BETTER THAN YOUR LAST

You have a duty and obligation as a trainer to continually polish yourself, your lesson plans, your presentation (person/video) on a continued basis. Think of your profession as a necessity to feed your family. In the private sector where I work, it is simply that way. If I do not continually improve my classes, keep myself current technically, tactically, and with information, someone else will take my business and livelihood away. My family is too important for that. So, I work a little harder rather than a little less. It has never hurt me to maintain this philosophy or mind-set.

Think of the students as people who are going to feed their families on their talents. In the case of tactical training, think of students as individuals who are going to war tomorrow. Your training is either going to keep them and their buddies alive or it is going to allow them to die. This thought may help motivate you to become a better instructor. You will look for more innovative ways to get the information into their minds.

## THE HARD CASES

Occasionally, the civilian world has a "hard case" to deal with. I define a hard case as one of those unpleasant incidents where someone is hurt or killed as a result of training, daily work, or a mission. The law enforcement and military arena provide the easiest examples to point out. For example, a person can be injured or killed as a result of friendly fire during a real-world mission or training exercise.

Generally the consequence of such an incident is that all training is ceased, interviews begin, and statements are taken. At the incident level, where a specific team is involved, usually one of two-type teams is present. One is the team who has been together for a long time through thick and thin. This team sticks together on their story and generally point out what happened and who is at fault. If they are a professional team, the team member who screwed-up and demonstrated the poor judgment will stand up and admit fault by describing how he shot when he should have held his fire. Much depends on the atmosphere that leaders of the group have built over time.

If leaders always preach and demonstrate honesty, integrity, and candor, then the men will step forward without hesitation. Occasionally, the professional team will try and protect the individual by altering the story a bit by leaving out a detail here and there. This can cause problems when future problems arise and they will continue their pattern of cover-up.

The other team that can be involved in the incident is the team that has been newly formed or has self-serving leadership. Depending on the professionalism of the leadership and the guts of the TL, the team may or may not bring the proper information to light. The new TL can be pressured by team members and help the "good guy" who was involved. The incident may be covered up. Once in a while, a TL will stand up as a leader, call a spade a spade, and set a new standard for the team. This may be difficult to do, but it is the right answer. A good leader and instructor can turn a negative experience into a positive learning and building experience.

## GOLDEN RULES OF TACTICAL INSTRUCTION

If it hasn't already been made clear by now, leadership and tactical efficacy are linked together. One cannot exist without the other. Understanding the concept of leadership is the first step to creating and sustaining any effective fighting force. Now that I've examined the dynamics of leadership through my own special ops experiences, I want to offer more a more detailed description of the tactics and techniques that I have found to be vital in any combat situation.

As you examine the tactics covered and reflect on your own tactical and training goals, keep in mind the fact that any teacher must be a student first. Understanding effective leadership skills and tactical training techniques is completely useless if you cannot share that knowledge with your team members.

---

### Golden Rules of Tactical Instruction

- Never close a student's mind.
- Leave your *ego* at the door.
- Prepare for each class.
- Don't be afraid to say, "I don't know."
- If you don't know, *ask*.
- Fix two items every class and make the next class better.
- Explain, train, rehearse, evaluate, critique, and rehearse.
- Don't say this is the only way to do things.
- Believe in the tactics and techniques you use.
- Keep yourself ready for the fight.
- Treat each teaching experience as if your students were going to combat tomorrow.

---

## KEY POINTS

- List the reasons why you want to teach, then let someone in the same field look at them and get an opinion as to whether you're teaching for the right reasons and not just for a paycheck.
- Lose the ego. It will cause you too many problems and close too many doors and minds. Learn to say, "I don't know."
- Record what works for you and what does not. Keep a list.
- Make the next class better than the last. Some improvements will be minor, some will be major. If you improve each class, within a year, it will only need minor polishing to ensure you are on the cutting edge.

# 13

# UNDERSTANDING THE STUDENT

*The mind can only absorb, what the ass can endure.*

—From an unknown, but wise, professor

- **STUDENTS—DIFFERENT PERSONALITY TYPES**
- **BACKGROUND OF STUDENTS**
- **STRUCTURING FOR THE COMMON STUDENT**
- **EXPERIENCE AND ATTENTION SPAN**
- **DEALING WITH PROBLEM STUDENTS**

## SCENARIO: FIRST TIME IN THE BIG D

I remember teaching my first Advanced Hostage Rescue class to one of the two Dallas tactical teams in the early days of this century. I had limited PowerPoint material and a dry-erase board. None of my drills had yet been videoed or imbedded into the PowerPoint, I had to explain the drills via the dry-erase-board. I had to walk the students through the drills, and then allow them to rehearse.

It was a tough crowd, as many of the veteran officers had literally thousands of raids under their belts as part of their team. This was the pre–*Dallas SWAT* TV show version of the team. The team commander, Bob, had brought me in to teach multiple-breach-point operations. A recent police raid in Cobb County, Georgia, had left two officers dead partly because of a single breach point assault. Bob wanted his folks to be able to perform multi-breach-point operations without fratricide.

I was teaching the first day's material. I covered exterior movement, how to get the team to the breach point under a variety of situations such as gunfire, runners, unknown individuals, etc. During my lectures, I got that "Who the hell is this guy?" impression from some of the guys. Some members believed they had tactics that were successful and were comfortable with those tactics. I suggested that we try techniques, theirs and mine, and see, "who got shot less."

I used video recording as my primary tool to record what happened during each scenario, and then I showed them the results. I have found that when someone is put under pressure, their short-term memory can only absorb so much before valuable data is lost. In short, I avoid calling a student a liar and put him on the defensive. Instead, I will say, "Let's watch the video," and allow them to make up their own minds.

I recorded each team responding to five situations, then allowed the teams to watch and critique their actions amongst themselves. Instead of reviewing the video immediately after the exercise, I would wait until the next morning when everyone was rested and fresh.

My first question would be, "How hard do you want me to critique?" Most would say, "Hard." I would get my laser pointer out and begin the critique. I would look for "gross" or safety issues first. Items such as pointing a weapon at your buddy or moving in front of a shooter who had priority of shot were addressed. I pointed out the finer details. I would never demean a student, but rather point out, "These are your options," or ask, "How can you fix this problem?"

In the end, I did not put down the team's tactics or techniques, but allowed the officers to see both systems and choose for themselves. I also used team members as role players in scripted scenarios against other teams allowing their own peers to the critique. This helped as their peers would describe the speed of the officers and their lack of ability to aggress the bad guy. This helped make believers out of skeptical officers without bruising egos or creating ill will. Also, by watching the video, officers can see real time what is happening, even if they were out of the action or in the back of the stack.

---

## STUDENTS—DIFFERENT PERSONALITY TYPES

Many of the students I run across in the tactical community are type A personalities. They are highly motivated, dedicated, and driven for self and unit improvement to the point of perfection. They are sharp, above-average performers who put in the extra time to ensure they are a cut above the rest of the crowd. When dealing with type As, one must remember that they have feelings, egos, personalities, etc., and can be turned on or off with a word or gesture. While you are assessing them as students, they are assessing you as an instructor. Expect this and do not get defensive.

Students will range from older and experienced to younger with limited or no life experiences. Keep all this in mind while teaching. Every student you come in contact with is on a different journey in life and at a different point. Be understanding of this.

## BACKGROUND OF STUDENTS

In the law enforcement tactical community, SWAT officers come generally from the ranks of patrol. These are individuals who want more, who want to be tested and pushed to their limits. The problem is looking a bit farther back. Where do the patrol officers come from and what are their experiences? Some agencies require a college degree. Some allow military service as a substitute. Look farther back and ask, how was their high school experience and young adulthood, and how did it shape, or fail to shape, their lives?

In general, I have also witnessed three types of students:

- Ordered to come to training
- Volunteered to come to training
- Paid their own way to training

The student that is ordered to come to training can be either positive or negative about the experience. I find the most negative experiences take place with students who do not wish to be there. Some are curious about what you have to say and will listen and take it back to their departments. You have a chance to reach these students and instill a passion for the profession and work ethic. Students who have volunteered to attend training wish to be there and further their knowledge on the subject, and those students are generally positive in the class. Some attend for the wrong reasons and wish to pad their resume rather than do a good job or give back. Either way, they want to be in the class.

Finally, the most intense and motivated students are the ones who pay their own way. These students will be on time, take notes, ask questions, and for the most part, be ready to train.

## STRUCTURING FOR THE COMMON STUDENT

We have discussed that an instructor needs to be tactically proficient in a skill set, and this is critical to understanding mechanics. The next big hurdle an instructor faces is how to break down that information and relay it to a student so they can understand it. Below is a student handout that contains all the mechanics on how to make one shot.

## Anatomy of a Tactical Shot

| | |
|---|---|
| Holster | Used in day-to-day travel. You can access the holster from the top or the sweep, using the holster as an index point. The weapon is drawn up and into the high ready position. The trigger finger only goes into the trigger guard when the decision to terminate life is made. |
| Depressed Ready | DR (or *sul*—Portuguese for "south") is used when in the stack, when another friendly is in front of you or in a relaxed state. Handgun is rolled over support hand and indexed on the support index finger, weapon pointed at a forty-five-degree angle toward the ground maintaining a straight trigger finger. |
| High Ready | Aggressive fighting position used when no friendly officers are in front of you, generally in CQB or closing on a threat. Elbows into sides, weapon centered under shooting eye, muzzle slightly elevated to catch front sight first in peripheral vision when the weapon is pushed into the firing position. |
| Vision/Scanning | Vision goes from a soft (whole person focus/big picture) to specific threat (weapon) and intent. Once the decision is made to terminate life, the eye then |

| | |
|---|---|
| | picks a spot on target and gun is driven to this spot using the front sight, which is slightly elevated above the rear notch. |
| Presentation | When you decide to terminate the threat, the weapon is driven/pushed forward and arms are locked out to the spot the eye has selected on the target for the front sight to arrive on. |
| Sight Alignment | Front sight is pulled down into the rear sight or held slightly above depending on surgical necessity of the shot required and the distance to the target. |
| Grip | Grip should transition from calm to firm during this presentation phase and not become so tight it causes tremors. Use the line drill to help determine grip pressure. |
| Trigger Squeeze | Trigger slack should be taken 2/3 out during presentation and then fully taken out once the sights are aligned, the grip is now firmed up and the arms locked. Using a Glock trigger as an example, I take the slack out until I hit the "bump" or meet resistance. |
| Follow Through | This is the act of realigning your sights and taking the slack out of the trigger once the first shot is fired. This is the fastest way to get your sights back on target and your gun ready to fire should the threat still be active. I use the long Glock trigger pull to help pull the gun back on target and take the slack out/reset trigger. |

| | |
|---|---|
| Cover | Once the target has fallen from your sights, you scan right and left of the target and then to the ground (where the threat should be). The weapon should be below the line of sight and the *trigger finger should be straight* during this phase. Another technique is following the threat to the ground and then looking right or left. Both work. I prefer the first method for multiple opponents and for a soldier on a battlefield. |
| Scan | Bring the weapon into the high ready and scan right/left to your rear, over your shoulders, keeping the weapon down range with a straight trigger finger. You are looking for fellow officers, their positions, and their weapons demeanor before you move. As a civilian, you are looking for other civilians or a police officer that might have arrived and who might potentially see you as a threat. |
| Close/Cover/Move to Cover/Retreat | You can now *close, cover, move to cover*, or *retreat* from the area. |

The box above provides a brief description of what goes into the firing of one single shot. Once I started teaching, I learned to reverse engineer or explain the mechanics to a student in a simple fashion. I quickly found out that this explanation can be more detailed and complicated than it seems.

I have a friend who is an exceptional martial arts instructor. Using this analogy, I asked him to explain how he would teach me to make one punch using the same concept. His brain started churning. Many instructors think it is only a simple punch. Reality is that there are many points that go into this equation. A punch engages multiple elements of a body's mechanics, such as the upper body, arm, and hand tension; how fast to drive; what parts of the body move to help generate power; when to tense up the hand; and how to tense it prior to the strike and the kind of follow-through necessary to achieve maximum penetration. To put down all the finer points, it would easily take a page and a half to two pages of written instructions.

What separates mediocre instructors and great instructors is this attention to detail and the conveyance of detail to a student in a manner easy to understand; this is a challenge for new instructors. The new instructors may be world-class shooters, but unless they can teach me how to get from point A to point B in an efficient manner, they are simply mediocre instructors and are not making the best use of my time. It is easy to say, "Watch me shoot," and then say, "Do it like I did it." This is a poor method to get the information across. All instructors are guilty of this at one time or another. The instructor must then go back and answer all the little questions that would inevitably come up. Receiving only a few questions means I delivered a complete and organized block of instruction.

When teaching tactical shooting on a range or in a field environment, I try and give students a mental and physical break after every three magazines of fire, whether it be with the rifle or pistol. I teach a one-shot procedure with several steps to it. If students follow it for each shot, they will be mentally exhausted before they are physically exhausted. Also, most students who show up are not in the physical shape required to train hard, so I back off on the intensity and promote an enduring training session that will tire, but not exhaust them. I expect this as many do not wear their Kevlar vests and helmets except on call-outs.

As for the mental aspect, I try and show them where they are at, where they need to be, and how to get there. I teach them how to train on their own once the class is over and how to honestly test themselves. Too many times, students fire thousands of rounds during a class and, when it is over, say, "That was a blast." Yet afterward those same students have little or no useful training to take back with them.

I have found that an intense range session can last about one and a half to two hours before a sit-down break is needed. Normally, I tell students to drink water and use the facilities during the time I give them to load magazines. They can decompress and chit-chat with fellow students at the same time, but they need to get back on line. As an instructor, when I see them fatigue mentally or physically, it is time for a longer break. I may throw in subclasses to divert and change their attention.

Classroom is much the same. For the type A personalities, I try and only do half a day in the classroom, then take them out for field training or hands-on the other half. I occasionally sit on the other side of the table as a student to remember the pains of sitting in a class for eight hours. As for classroom breaks, I try and give breaks every hour and throw in funny videos to break the monotony. This is especially important after lunch when the infamous "food comas" set in from overgrazing.

## EXPERIENCE AND ATTENTION SPAN

As a trainer, I teach different types of classes. Some will be high risk. Some will be low risk and some will be technical or lecture. During high-risk training, it is important to bring all students to the highest-level safety. This is critical because "one aw-shit wipes away all the attaboys." Besides, it is unprofessional when an instructor injures a student out of negligence. The instructor might as well kiss his reputation good-bye. The word of mouth in the law enforcement arena is as fast as any I have witnessed. Students have called friends and coworkers each day to report the progress, likes, and dislikes in classes. That word multiplies back at the agencies, the next day, as that officer tells their coworkers.

I have also witnessed the posting of class AARs on the chat forum boards the same day a class was over. It can be considered a report card for other prospective students. Screw up and a great deal of folks will hear about it. Do a good job and it will help your business grow.

I am talking about experience and attention span, and it is the instructor's job to craft a lesson plan/class in a manner that looks out for students' experience, trains them in the basic skills, then brings them all on line at the same time. With high-risk training, all students need to possess basic safety skills with rifle and pistol or they cannot proceed with live fire.

Review and teach them the safety skills. I demonstrate it, allow them time to practice the skills, then remind them when they are not doing a particular skill correctly. It is critical that all personnel be safe because one negligent discharge can cripple, kill, or ruin a career.

I have a safety valve of sorts when conducting high-risk live-fire training. That is to issue paint-marking weapons when students cannot grasp or execute the safety fundamentals due to experience, past training, etc. I will cover this more in detail later in this book. Attention span in class is different from attention span in high-risk training and should be constantly monitored and observed. In class, you can see students' minds wander as they become bored or restless, then you know it is time for a break.

In the tactical training business, the same applies. I know the signs indicating a student is losing focus on the task at hand or is mentally exhausted. It has been said that the last ski run of the day is when most students get hurt. They are tired and push themselves too far and wind up in a cast. The same applies to high-risk training and especially CQB. If I push students too far or too fast, I am asking for an accident.

## DEALING WITH PROBLEM STUDENTS

### *Unmotivated, Lazy, or Lethargic*

There are several ways to deal with unmotivated students. Many times I will not have time as an instructor to find out what is the in-depth problem with an unmotivated student. My job is to set the example and be a positive and quiet professional and set the tone for the class to follow. I task my students to help ensure the class runs smoothly. My job is to run the class, not let unmotivated individuals drag the class down.

One technique is to assign the unmotivated students to a positive and upbeat team with a strong TL and let them see what it is like to be on a good team. Let the team create a positive family environment, and hopefully, the individual will see the light. If they are unmotivated to a detrimental effect, they should be counseled by the lead instructor and the team should take a neutral stance on having them on the team. The instructor should not harass, but get the needed performance out of the individual, then ignore them. It will make for a slightly uncomfortable environment, but sometimes in training you get individuals who are simply getting by trying to steal others' oxygen.

### *Disruptive*

*I hate able-bodied, arrogant, closed-minded individuals who stifle their own growth and that of those around them.* I will not tolerate disruptive behavior in a class. I will talk to the person during a break. If this does not correct the behavior, I will remove them from the class. Poor behavior will not only distract from the class and other students, but it will also piss me off and cause me to miss valuable teaching points. Once they are removed, document it and have other students in the class write a short narrative as to what they witnessed. If I feel it is warranted, I copy the packet and send it to the chief where the individual works.

### *Inexperienced or Ignorant*

One definition of ignorant that I was taught is "where the truth is lacking." By this definition, many individuals do not have the experience, knowledge, and background necessary to ask questions. That, coupled with peer pressure, may inhibit them from asking questions in front of the class. They may be more inclined to ask questions in small groups within the team concept that I will describe later in the book.

For these individuals, I have simple and easy-to-understand lesson plans that are comprehensive and answer most of their questions up front. When possible, take inexperienced/ignorant individuals and assign them to strong "battle buddies," or partners who will help them through the class. Use all the talent at your disposal and allow them to help develop the weak links. Remember, you have teachers among the class and do not be afraid to use them.

## KEY POINTS

- Understand students and be patient with them. Realize they come to you with varied backgrounds, experiences, and motivations. Your job is to funnel them into a cohesive class and facilitate a positive learning experience.
- Structure your class for the common student.
- Understand the length of attention spans for both technical and high-risk classes.
- Learn to handle disruptive students. Do this quickly and efficiently.

# 14

# UNDERSTANDING THE WHOLE

*The whole elephant or the toe, which makes more sense when you describe the animal?*

- **STRUCTURING A CLASS AND UNDERSTANDING THE WHOLE**
- **CLASS ADMINISTRATIVE POINTS AND WHERE WE ARE GOING**
- **HOW WE ARE GOING TO GET THERE**
- **GENERAL STRUCTURE**
- **SPECIFIC STRUCTURE/MODULES**

## STRUCTURING A CLASS AND UNDERSTANDING THE WHOLE

For most of us, our minds need a flowing structure and development to gain information from a subject. This can be done in several ways. My approach is to have the instructor review the lesson plans and ensure the material flows logically and builds on previous lessons. This is a critical part of the instructor's profession. Instructors should take the time to structure a class so it flows and each module builds on the one before it. If the instructor does not take the time to do this, then the student must take valuable time to process disjointed information, constantly going back and referencing other material. Valuable information will be lost if the student cannot process the information effectively.

Instructors must occasionally update new information and insert it in the appropriate logical sequence to keep the mind open and flowing rather than clog it up with sporadic, disjointed information. I relate it to performing a disjointed physical training program in the gym. If your weight training system does not build and complement on each move or exercise, your body will not get the most out of the training session. This same concept applies to structuring classes.

I believe it is critical to keep a student's mind open. Seeing the big picture in the beginning is critical to the student as well as to the instructor. The instructor should constantly review the basics and understand where the class is going and how to get there.

For the students, seeing just the "toe of the elephant" does not give them a picture of the animal. The student must have a point of reference (big picture) to constantly look back on to see how the material relates and how important it is to the whole.

## CLASS ADMINISTRATIVE POINTS AND WHERE WE ARE GOING

Students will make faster and more efficient strides in learning if they understand the big picture and how they are going to get there. Using my Advanced Hostage Rescue class as a sample model, I will explain how to answer students' questions and put their minds at ease.

I give a brief introduction to the class, including who I am and my background. I have each student talk and give a brief bio. This gives them the chance to speak to the class and gives me an idea of who they are, their experience level, size of their department, special skills and training, and finally, any leadership or management experience.

While the students are giving their bios, I take mental and sometimes written notes. I try and capture who are my leaders and start mentally assigning TLs. Why? They are a force multiplier for control, information, and leadership. In the end, I can give instructions to four students (team leaders), assign tasks, request help, and generally practice a much-needed leadership system that works and may not be in place in their home department.

Next, I break the students down into teams and assign team leaders. After that, I take team pictures to put on a consolidated sheet of paper, so I can learn first names in the next day or two. If by day 4 or day 5 I call a student "sir" or "ma'am," it is because I don't know their first name. I keep sheets for when I travel back to the same areas for conferences. I study the photos so that way I can remember their first names when I see them again. I like to think it shows that I care and usually makes the student feel good about the encounter.

We take a break after the photos and let the TLs meet and talk to their new teams. I have TLs get contact information and phone numbers from their team members. I want TLs to be accountable for their people, and if someone is going to be late to class, I want the TL to tell me first. I effectively shorten my span of control in the class to four people. I hold TLs accountable for team members, safety, training, etc., again, in an effort to implant, demonstrate, and practice a proven and effective leadership system that transcends the classroom into the tactical world.

## HOW WE ARE GOING TO GET THERE

Using the training schedule in the student manual, I discuss the course flow for the week. I take each day and break it down into sections and discuss what is to be covered in the blocks of information. I talk about why each day is structured the way it is and why it builds to a culmination for the week.

## GENERAL STRUCTURE

Using the below "cheat sheet," I will elaborate on why I structure a class the way I do. The outline is my daily instructor's checklist to ensure I cover all the information in a logical sequence.

---

*DAY 1: INTRO/ADMIN/HOSTAGE RESCUE OVERVIEW/ EXTERIOR MOVT*

- My Intro
- Group Introduction
- Roster
- Team Breakdown
- Team Photos
- *BREAK*
- Safety—Paintball/Vehicles — POWERPOINT
- HR Cycle — POWERPOINT/COBB COUNTY
- *BREAK*
- Combat Mind-set — POWERPOINT
- Equipment
- Medical Kit

- *LUNCH*
- Weapon Clearing Procedures
- Stack—knee
- Movement/Formations (Flare) VIDEO/POWERPOINT
- Actions on Contact
- Suspect Manipulation
- Officer Down
- Scenarios VIDEO

*CLASS PREP*

- Safety Letter
- Personnel Roster
- Business Cards
- Packets
- Assault Gear and Rifles
- Medical Gear
- Classroom Box
- PowerPoint Equipment

---

I break day 1 down into classroom and field time so as not to keep a student seated for eight straight hours. I begin with an overview of the week. My first class is safety. I have seen too many training accidents over the years where students have been killed by other students, and even instructors, during training. I set the stage for training in this class from day 1 and tell them what is expected and why.

I give the students an overview of hostage rescue and possible variables they may encounter. I show, through video and PowerPoint situations, what has happened in the past. I also show schoolbook answers to the "generic" hostage rescue scenario, and then build on them through the week to common core situations that have happened in the past.

I discuss combat mind-set (for more detail, see chapter 1), what I expect from them and how to develop it. I want students to fight through and solve every problem without giving up. I have found that if you have students practice "dying" or stopping when hit in training, they will do the same in a tactical situation.

After lunch I teach a short block on exterior movement, which deals with how to get from a drop-off point on foot or from a vehicle to the breach point of their target. I show what scenarios I want the students to prepare for and allow them to practice the scenarios with their teams. I use this short training block as a sample to demonstrate the training the rest of the week. This includes weapon safety checks, team rehearsals, video of scenarios with paint-marking rounds, and AARs.

---

*DAY 2: CQB*

- Review Day 1 Video
- *BREAK*
- Safety
- CQB — POWERPOINT/VIDEO
- Flash Bangs — POWERPOINT/VIDEO
- Assault Breaching — POWERPOINT/VIDEO
- *Lunch*
- Scanning Drill — VIDEO
- Flash Bang Drill
- Hostage/Suspect Manipulation Exercises
- Closing Down a Room
- Force on Force
- Medical Emergencies (Team Demo)

*CLASS PREP*

- Hands/Guns

- Individual CQB Cheat Sheets
- Pistol Standards
- Pants/Medical Stuff
- Breaching tools
- Rifle
- Targets
- Classroom Box

---

On day 2, I begin with a review of the video from day 1 including exterior movement. Using my laser pointer, I offer one to two solutions on how to fix any noted problems. This keeps them in a positive mind-set of looking for and implementing solutions, not cringing from an instructor's scathing comments. At the conclusion of the video review, I ask the class to give me three to five points that we can polish up should we see this drill later on during the week. This way, I check to see what the students learned from the video review and give them a chance to improve.

Day 2 consists of CQB and flash bangs or noise flash diversionary devices (NFDDs). I show students the different systems of CQB and explain why I use and prefer my system. I move on to the employment of flash bangs and show a PowerPoint class loaded with video of how to do it and bloopers of how not to do it. Later, I issue "dummy" flash bangs for practice during the rest of the course. I have found that if students do not practice employing flash bangs, they will carry them through an entire target on their vest.

This is a busy day for me. After the classroom lecture, I head straight over to the training site and set up targets for a discrimination room. I usually put fifty targets in an average-sized room and have students come in with two-person elements and conduct a two-person scanning exercise where they visually scan for two weapons in a sea of targets. Each target has two hands, some with props, some without. I capture the students' entry and movement with the video camera to show the next morning. I am looking for the proper mechanics of speed, movement, weapon posture, etc.

I record each element two times, then push them out to a training area to work on two-person and team-level entries. I want the TLs to get used to conducting rehearsals with their teams in a dry run. Again, another leadership challenge. While performing dry runs, the teams can practice opening the door, flash bang procedures, etc. If done correctly, the newly formed teams will learn how to work together to smoothly accomplish the scenarios I have injected them into.

Once all teams are videoed, I turn the scanning room over to the teams to practice on their own with two-person and team elements. I take two teams, having one strip their gear off and wear civilian jackets, to act as role players inside a room. I then have the other team come in and practice putting hands on live bodies, flex-tying with deactivated flex ties, moving and searching people.

I use a checklist of common situations encountered in a room and give students the answers to the problems. They will face these same scenarios later on during the week. After both teams have been through this module, I take the entire group and run a medical exercise. I take one officer and issue them a pair of pants. I have one team come into an area where they encounter an "unknown" suspect and decide how they should deal with them. As soon as they give an "all secure," I start squirting blood on the patient's leg injury and require the officers to perform their security protocols first, then handle the patient. I have them practice it twice, then ask if there are any questions. At the conclusion, I issue bandages to each team, clean up, and tell them when and where to report tomorrow.

Medical scenarios are probably the least practiced drills I have encountered. Generally, officers are taking twenty to twenty-five armed individuals into a house against armed bad guys and no one is carrying bandages. I push for everyone to carry bandages because they are the first responders. They can get shot, their partner can get shot, innocent civilians can get shot. They will be able to provide the most rapid life-saving medical aid in most situations. Providing aid to their bodies, should they be shot, is critical to a positive combat mind-set and personal survival.

---

*DAY 3: INTERIOR MOVEMENT*

- Review Video Day 2
- *BREAK*
- Safety
- Interior Movement — POWERPOINT/VIDEO
- Corner Clear
- T Intersections
- Hallways—Single
- Hallways—Double
- Breaching Setup
- Stairs/Elevators/Restrooms
- Officer Down

*CLASS PREP*

- Breaching tools

---

As with day 2, I begin day 3 with video of the CQB scanning exercise and, again, ask them how hard to critique. Once I make a couple of points, I do not keep beating them up. I know they get the picture.

Our core class of day 3 is Interior Movement. I previously taught the students how to move from a drop-off point to a breach point and how to process rooms with CQB techniques. Now, I have to show them how to negotiate hallways and get to the rooms or crisis points. As with exterior movement, I show the students PowerPoint of how to do all the drills from open-door, closed-door approaches to single- and double-side hallways. The key to this day is to teach how to solve a single problem with a single team and not to overwhelm with too much information. If each team can solve a problem and move on the offensive again, then multiple teams can take a great deal of ground in a hallway in a simple and efficient manner. As with other training sessions, I video the drills from the front and from the rear for review. I use five common core scenarios that I want the students to learn.

I begin with five runs per time as a single element, then move to two team movement in the hallway. This usually takes several hours, but provides great learning video. Each team will run the same series of scenarios choreographed by role players who are instructed not to deviate from the script. It is crucial to control role players as they will want to "win" and sometimes they forget that the goal is to train.

---

### *DAY 4: TEAM CQB/PLANNING*

- Review Video Day 3 — 2 Plus Hours
- *BREAK*
- Safety
- Assault Planning — POWERPOINT/VIDEO
- Team Leader Duties
- Assistant Team Leader (ATL) Duties
- Sniper Responsibilities
- Individual Responsibilities

- Stealth versus Dynamic
- Known versus Unknown Location
- Multibreach Linkup and Coordination Points
- Antifratricide Teaching Points
- Hostage Evacuation
- Medical Emergencies
- Improvised Explosive Devices (IEDs)
- Assault Breaching Video
- Assaults
- Team Planning PE

*REHEARSALS*
- Simultaneous MultiBreach
- Deliberate Assault
- Stealth Entry
- Compromise
- Door Breaching
- Window Breaching
- Flash Bangs

*CLASS PREP*
- Classroom Box
- Dry-Erase Board
- Target Folders

---

I begin this day with video review from the front, which usually takes an hour, then review the video from the rear, for a total of two hours. A great deal of action takes place during these scenarios and students do not get to see the entire picture because the team may have been split, some in the room and some in the hallway during the action. The video gives them a chance to see what really happened and at what speed. I usually call this portion the "second half of the hallway movement class." Again, I show this first thing in the morning because the students are fresh and rested and will better absorb the material and critiques.

Students will get to see, real time, how long they stayed in the hall and how often they exposed themselves. They also see how efficiently they processed bad guys, injured hostages, etc. They will see if they dropped security at any time. This is generally an eye-opener.

After this review, I put the entire package together for the week. To this point, the class has performed individual team modules. Now that the teams know what to do, I need to be able to put multiple teams on a target without killing each other. I teach planning classes, including how to plan a multibreach point target. I have TLs come up first and perform tabletop exercises and brief their plans. Then, I have their ATLs come up and plan in case the TL is missing. I want everyone to understand the leader's role in planning, how to plan breach points, and how their actions relate to other teams in the mission.

This day is a classroom day, all day. I break up the class into working groups and teams to develop target plans on different targets and how they would assault those targets. I have TLs brief the deliberate plan and the ATL brief the hasty plan. Everyone can ask questions on the various plans. This teaches TLs and ATLs how to brief and what I deem is important. I also use various targets so students start developing "template" plans for different structures encountered in their communities. Students learn the importance of numbers and how many they actually need to execute a successful hit versus how many they have. Students learn to give multiple tasks to one team so that they can accomplish more with less. Students also figure out how many breaching tools their team needs versus how many tools they have. Finally, students learn a planning sequence that they will employ and reinforce the next day during the final training exercises. This sequence can then be taken back to their department and used on real world missions.

---

*DAY 5: SCENARIOS*

- Deliberate
- Deliberate with Compromise
- Stealth
- Deliberate with IED
- Emergency Assault X4

---

Day 5 is a scenario-based day. The class practices the planning process on how to brief missions. The class conducts multibreach point exercises with paint guns. I do not record the scenarios with video cameras as it takes too long to debrief and I would lose momentum. Instead, I conduct AARs using a structured format. I do not care who went right or left during the scenarios, but what problems were encountered and how they intended to fix them.

I emphasize that if they fix two problems per exercise, they will be doing great at the end of the day. Usually, the first hit is slow, disjointed, and problematic. Students hit the same target again, using the same scenario, and fix most of the problems they encountered on the first run. As the students gain confidence and begin to perform better, I add problem-solving modules. I might add a second shooter, a medical emergency, and an officer-down scenario and have the students work through the problem. I preach to them that all they need is to solve one problem at a time and move on to seize more ground. If everyone follows my instructions, it will run extremely smooth.

At the end of the course, I recapture the week and what was accomplished from day 1 and give my thanks and appreciation for their service to our country.

As a general rule, I strive for 20 percent classroom to 80 percent hands-on during the course. Starting with day 1, I try to front-load all the boring "admin classes." Some class structures have required classes or modules that need to be taught as a certification requirement. I push most of that information on day 1. As the week goes on, it is not their physical stamina that erodes, but their mental stamina.

I also try and knock out my classroom portion in the beginning of the day, then transition to hands-on modules the remainder of the day. I use the administrative classes to explain to the students the reasoning behind the drill or exercise and how it relates to their survival. When put in this context, students are more likely to accept it.

## SPECIFIC STRUCTURE/MODULES

I want to pass along a couple of thoughts on how I structure a module or short block of instruction in a class. The classroom portion of the lecture explains the system to be used and why. I try to teach a common system that uses a slight variable for different scenarios encountered. Once there are no more questions during the classroom portion, we move to the training phase.

Student TLs are responsible for organizing and accounting for their teams, equipment, and safety checks. Once the teams are checked and assembled, a walk-through demonstration of the drill is performed with one team while the other teams watch. After questions are fielded, teams are assigned to different training areas and TLs take charge of their teams and conduct rehearsals. Once TLs are comfortable that their team can execute the drill in a safe manner, they can be assigned a role player to now practice the drill with an interactive component.

The team practices for a short time, then all the teams come together and a safety plan is briefed. The teams prepare for video-recorded short-duration scenarios where the teams can practice the drills under simulated gunfire stress. Each team runs several scenarios back to back and a short debrief is conducted after each run. At the conclusion of the runs, teams are allowed to take a break and can dry-run anything they had problems with in another training area. Shortly after this, two teams are injected into a set of common scenarios, with video, to see the problems encountered with multiteam movement.

At the end of the day, the teams are debriefed and told to report to the classroom in the morning to watch the video. Again, the reason for watching video in the mornings is because the students are fresh, not worn out, and can mentally capture learning points more effectively.

---

At the end of the week, students have learned how to execute the following:

1. Move to a breach point from the outside and handle any problems encountered
2. Handle breach points
3. Move down hallways as a single team and as part of an assault force
4. Conduct two-person CQB and team-level CQB
5. Conduct planning and multi-breach-point assaults on structures, effectively secure and orderly evacuate the same with a host of scenarios

---

## KEY POINTS

- Structure the class flow so that it builds and reinforces previous lesson plans.
- Always use internal leadership during training and rehearsals to ensure students know the correct way to lead and train. Students cannot get enough of this.
- Review video in a productive and rested environment. Do not close a student's mind.
- Tie the course up with a final exercise(s) that reinforces *all* the lessons from the entire week.

# 15

# DEVELOPING TOOLS TO TEACH A CLASS

*If you can develop one class, you can develop a thousand. The formula remains the same.*

- **GOALS/OBJECTIVES FOR THE CLASS**
- **TARGET AUDIENCE**
- **LENGTH**
- **INSTRUCTOR TO STUDENT RATIO**
- **TESTING**
- **REFINING THE TRAINING**
- **CREATING A POSITIVE LEARNING ENVIRONMENT**
- **STRESS IS WHAT YOU PERCEIVE IT TO BE**

## GOALS/OBJECTIVES FOR THE CLASS

Goals should be simple, easy to outline and understand by the student and other instructors who may inherit or teach your material. Remember that you are developing information delivery systems for students to assimilate and for other instructors to deliver. Both sides need to be able to understand it. Instructors need to teach under the individual who developed the information. Otherwise, information gaps may occur. One of the best systems I observed had a year overlap before the current instructor handed his material to the assistant instructor. Once this happened, another assistant instructor was assigned and the overlap process started all over again.

### *Familiarize*

Because of the amount of material I teach in my courses and the amount of time it requires to master, I familiarize students with my systems instead of trying to test or ingrain the material.

As noted earlier, current theory states that it takes two thousand to three thousand repetitions to master a physical act or develop physical "muscle memory." I believe it takes an equal amount of "mental muscle memory" to ingrain a thought process or act. It will take a student more than one day or a week to master tasks and skills that have taken me years to develop and perfect.

As an instructor, I show students how to perform the task. The students must mentally and physically take notes to integrate it into their personal training system. Over a period of time, this is the only way they will "get it." I understand this as an instructor when I write the programs and when I give students more information than their sponge (brain) can hold. Therefore in training, I cover the basics and core skills. I expect the students to implement the tactic or technique into their training regimen. This is the system I use for two-day introductory modules such as a Tactical Rifle or the Pistol One course.

## *Ingrain*

Ingraining tactics or techniques takes much more time to implement. I remind students that when I first went through Special Operations training, I dry-fired (dry practiced) with my weapon for eight hours a day for two weeks before I was allowed to fire my first shot. The military had the time allotted for the training, unlike law enforcement training programs that are only a week in duration. In the civilian training arena, students would scream at the amount of dry-fire I used to perform. Most students want to send bullets downrange to satisfy their idea of the fun side of shooting.

Because of their jobs, I have limited time with students. You must sometimes worry about how long to make a course (their shifts must be covered and someone must pick up their duties when they are away). I deliberately tailor courses for leaders to a shorter length because of the critical jobs many of them hold. As a side note, in reference to scheduling courses, I find it easier to run a course in a certain venue with a department or host agency should a crisis in the officer's area arise. He can leave the school and help his agency resolve it. This requires travel on my part. This also lets the chain of command sleep at night knowing that their people are in the area to respond to potential emergencies.

## *Shooting Standards*

I use *shooting standards* as a model to teach basic and advanced shooting classes. These standards or drills are a set of technical skills or basics a student must master to move to higher levels of training. They should be simple and comprehensive and apply to all aspects of the skill set.

My standards come from past and present training and from real-world combat missions. For example, I read about an incident where a civilian shot a bad guy who was wearing body armor. The bad guy spun around and shot the civilian, disabling the civilian's strong arm. Instead to transitioning to his weak hand and continuing the fight, the civilian took no action. The bad guy came over and executed the civilian. I am smart enough to know that if it can happen once, it can happen again. This is why I included support-hand-only fire in my basic drills.

The skills are organized in a way so when a student learns a drill, the next drill builds upon it through a natural training progression. I start with one shot on the pistol and break the one shot down into several subsections to help the student digest the information. I teach a "shooting checklist" that the students should mentally go through each time they prepare for a shot.

I start with how I plant my feet. I work my way up my body. I talk through the sequence and repeat all the problem areas that I personally encounter. For example, I say, "Right arm lock, left lock, tighten shoulders, firm grip, and press." This little sequence is what it takes for me to be consistent with each shot.

The standards are designed "gun neutral," which means they can be shot with any mainstream semiautomatic handgun. The times on some of the original standards are M1911 .45 ACP driven. The 1911-type pistols have a great trigger and are fast to reload. Most police officers will not shoot a tricked-out gun, but rather a department-issued weapon with a standard trigger. When I validate the standards, I shoot the same type of weapon as most patrol officers are issued to ensure that the times are attainable.

Next, I make sure each drill flows into street combat, vehicle assaults, or the shoot house without any changes. This way, when I teach a drill, it is a building block for success in other areas they don't even know about yet, and you do not have to duplicate effort or training. The drills are meant to be shot with an electronic timer to record times and add stress. The little beep of a shooting timer is incredible at creating stress in a student. It causes them to focus each time beyond the beep and on their skills. They must also be able to deliver the shot accurately in the kill zone under the time requirement.

By establishing a set of standards, I provide the students a comprehensive workout plan for the range. The question I ask students is that when they go to the gym, do they have a workout plan or do they just move around and sling steel here and there? All answer that they have a plan. Most are written down. If we do this for a simple physical workout, why not for a life-saving technical skill? If they can master the basics, there is not much more to it.

Occasionally, I will adjust the standards for new equipment that comes out that officers are required to use, such as holsters. New holsters with double or triple locking mechanisms may require an adjustment of the time. No matter how fast the officer is, two or three locking devices will slow down draw time. I adjust times as required.

---

## PISTOL INSTRUCTOR STANDARDS SEPT 2010

The drills below are designed with three purposes in mind:

1. A measurable standard to maintain
2. An efficient stair-stepped workout program that covers all the bases
3. To test the individual shooter at various times to show areas needing improvement

Only score shots in the center box and head of the CSAT target. If an enemy turns sideways, that will be all the shooter has to engage, resulting in a worst-case scenario.

| STANDARDS | | | | STANDARD |
|---|---|---|---|---|
| 1. Ready | 1 shot | 1 target | 7 yards | 1 sec. |
| 2. Holster | 1 shot | 1 target | 7 yards | 1.7 sec. |
| 3. Ready | 2 shots | 1 target | 7 yards | 1.5 sec. |
| 4. Ready | 5/1 shots | 1 target | 7 yards | 3 sec. |
| 5. Ready | 4 shots | 2x target | 7 yards | 3 sec. |
| 6. Ready | 4 shots | 2x weak/2x strong (1 target) | | 5 sec. |
| 7. Ready | 1 shot | Malfunction drill (1 target) | | 3 sec. |
| 8. Ready | 4 shots | 2 Reload 2 (1 target) | | 5 sec. |
| 9. Rifle up | 1 shot | Dry fire/transition | | 3.25 sec. |
| 10. Holster | 1 shot | Kneeling (1 target) 25 yards | | 3.25 sec. |

INSTRUCTORS MUST PASS 8–10 STANDARDS IN ONE COURSE OF FIRE.

-All stations shot at 7 yards except #10.

## TRI RIFLE STANDARDS SEPT 2010

The enclosed drills are designed with three purposes in mind:

1. A measurable standard to maintain
2. An efficient stair-stepped workout program that covers all the bases
3. To test the individual shooter at various times to show areas needing improvement

| STANDARDS | | | | STANDARD |
|---|---|---|---|---|
| 1. Ready | 1 shot | 1 target | 7 yards | 1.00 sec. |
| 2. Ready | 2 shots | 1 target | 7 yards | 1.50 sec. |
| 3. Ready | 5/1 Shots | 1 target | 7 yards | 3.00 sec. |

| | | | |
|---|---|---|---|
| 4. Ready | 2 shots/2 targets | 7 yards | 3.00 sec. |
| 5. Ready | 1 Rifle /1 Pistol | 7 yards | 3.25 sec. |
| 6. Ready** | 5 shots/1 target | 100 yards/prone | 20 sec.* |
| 7. Ready** | 5 shots/1 target | 75 yards/kneel | 20 sec.* |
| 8. Ready** | 5 shots/1 target | 50 yards/kneel | 20 sec.* |
| 9. Ready** | 5 shots/1 target | 25 yards/stand | 8 sec. |

*Total: 34 Rounds Rifle/1 Pistol*

- All rounds must be accounted for on the CSAT Target.
- Student must pass 8 of 10 drills to be successful.
- Drill 5, one shot from rifle is fired and then transition to pistol. Must both be hit.

* 4 of 5 rounds must be in box

** Requires only a full caliber hit on target to receive a "go"

**SNIPER STANDARDS** **JUNE 09**

The enclosed drills are designed with three purposes in mind:

1. A measurable standard to maintain
2. An efficient stair-stepped workout program that covers all the bases
3. To test the individual shooter at various times to show areas needing improvement

| STANDARDS | | | | STANDARD |
|---|---|---|---|---|
| 1. Ready | 1 shot/Unsupported/Prone | Body | 100 yards | 7.00 sec. |
| 2. Ready | 1 shot/Bipod/Prone | Head | 100 yards | 7.00 sec. |
| 3. Ready | 1 Shot/Stand to Kneel | Body | 75 yards | 6.00 sec. |
| 4. Ready | 1 Shot/Prone to Kneel | Body | 75 yards | 7.00 sec. |
| 5. Ready | 1 Shot/Stand to Kneel | Body | 50 yards | 5.00 sec. |
| 6. Ready | 1 Shot/Prone to Kneel | Body | 50 yards | 5.00 sec. |
| 7. Ready | 1 Shot/Stand to Kneel | Body | 25 yards | 4.00 sec. |

| | | | | |
|---|---|---|---|---|
| 8. Ready | 1 Shot/Prone to Kneel | Body | 25 yards | 5.00 sec. |
| 9. Ready | 1 Shot/Stand to Kneel | Head | 25 yards | 5.00 sec. |
| 10. Ready | 1 Shot/Low Ready | Body | 7 yards | 1.50 sec. |

*Total: 10 Rounds Rifle*

-Student must pass 8 of 10 drills to be successful.

STRUCTURE AND SEQUENCE

| | |
|---|---|
| • My Intro | |
| • Group Introduction | |
| • Team Breakdown/TL Responsibilities | |
| • Safety-Range Commands | |
| • Packets/Standards | |
| • Range Setup/Target Stands | |
| • Ready Position | Stance/Grip/Sights/Trigger/ Follow-Through/Cover |
| • Shooting Plan | Index Cards |
| • Dry-fire Sequence | |
| • 5 and 1 Ready Position | Dots |
| • 1 Shot Drills on Own | |
| • 1 Shot Drills on Timer | |
| • Line Drill/Trigger Reset | |
| • Reload | |

---

The above text is part of my instructor checklist that I use for a Tactical Pistol or Tactical Pistol Instructor class. After I complete the administrative and classroom portion, I move to the range. I have the student load three magazines of eight and report to the firing line with eye and ear protection.

I review a shooter's stance from the feet up, discussing weight distribution, positioning, etc. I move from the lower body to the upper body and discuss and demonstrate the hand/arm position for the high ready and the grip on the gun. I show students the proper load and unload sequence to use on the pistol. I demonstrate the one-shot sequence I use and the five-and-one drill. During this drill, I practice dry-firing five times correctly and I load the pistol and fire one live shot using the same sequence. I unload and dry-fire five more times and repeat the live shot for a total of five live rounds.

Once everyone understands the drill, students get on line and check their eye and ear protection before touching their guns. I conduct this the same way every time to keep a student from adjusting their ear protection with a loaded weapon. They are given the command to begin the drill and the line is "hot." They fire a total of five live rounds. This gives me time to walk the line and correct student errors without a great deal of ammunition being expended. Once the drill is performed, and the student has unloaded their weapon and placed it in their holster, they can take two steps back off the line. This way I can see who is still firing. I reiterate key points when everyone is finished and I check their targets. If they need more practice, I perform the drill again.

After this is accomplished, I introduce the students to the sound of the timer so as not to induce unnecessary stress. The "beep" will start them on future drills. I stress the need for proper mechanics over speed. My one-shot sequence is made up of several steps and it is critical to get this process down to establish a base for the rest of the course. Trigger reset and cover are the two hardest points to get a student to accomplish. Once they fire the shot, students want to immediately come off target and see where they hit. In a shooting or gunfight, it is critical to realign sights on the target and take the slack out of the trigger in case the bad person is still standing. It is the most efficient way to make a second or follow-up shot.

My next drill is called the Line drill. I have students shoot the left line of the target, starting at the top and working their way down with four shots, each time realigning their sights and preparing their trigger for another shot. They fire four shots, they fire four more starting back up. This drill tells the shooter about their grip pressure and trigger control. Griping too hard and pushing the shots left. The line drill tells the student if they are doing that. I repeat this drill several times until each student understands their own personal grip pressure.

After the line drill, I proceed immediately into how to reload a pistol with a magazine from a belt or vest. This helps build muscle memory early on and throughout the course. All these positive repetitions build muscle memory. Why wait until day 2 of a course to show them how to reload when they are performing this skill each and every time? They also have the chance to adjust their equipment to better conduct the reload.

By taking a little time and reviewing, the student is developing their own methodology. I can easily build a structured format that a student progresses and learns from in the shortest time possible.

## TARGET AUDIENCE

### *Basic Classes*

Basic classes start from scratch or a level playing field and bring everyone up at the same level, safely. Existing courses can be used as a model when developing a basic class. An instructor's experience shapes the class. The more current real-world experience an instructor has attained, the higher the level of preparedness the students learn.

Basic SWAT classes are a prime example. I start with shooting skills and simple safe movement drills with pistol and rifle to develop and reinforce skills to be used in the course. I start shooting on day 1, and every time students practice a drill, I reinforce the basics for other modules. Many students have varied backgrounds and experience levels and I must bring them to a common base for the class to proceed safely. In a short time, they are maneuvering around each other in tight confines with loaded weapons. Some students take longer than others to achieve a safe level, but in the end, all must demonstrate consistent proficiency to continue training.

I use the level of proficiency as a way to describe how high we can bring a student up in a course. "Operational level" is 100 percent ready to perform the job. For example, during my time as a Special Operations instructor, we used to try to get a student in the basic course to as high a level as possible. Reaching 75 percent of the level of a proficient Special Operation forces soldier was a realistic goal. Instructors realized that it will take a couple of years of practice, mentoring, and experience to be a solid stand-alone individual who is fully integrated into the team.

The lower the entrance requirements, the harder it will be for students to make the desired technical, tactical, mental, or physical levels. If selection standards are low, I might only get students to a 50 percent level when they leave the course as you can only go so far, so fast, with the talent I have (or don't have). Further, doing high-risk training, I can only move so far, so fast, without endangering the lives of other students with the bottom-feeders. As a result, in many law enforcement academies, standards are lowered to attain more applicants. With lower qualified applicants, academy staffs lower the standards and require less of the men and women who may have to fight for their lives in a high-risk situation the day they walk out of the academy. Lowering standards is doing the future officers a disservice.

### *Advanced Classes*

When conducting advanced classes for students with varying backgrounds and experiences, I still need to find common ground. I suggest a 50 percent review of basic material, and then add on 50 percent of new material. Problems I have witnessed in the tactical arena include different levels of weapon handling, marksmanship, and above all safety. This results from the various schools and so-called approved training that the state mandates. Not all schools are created equal, and some may be top-notch for many years and have their share of poor training years because of poor management practices.

I recommend starting advanced classes with the same safety briefing as a basic class and a review of firearms and weapon handling. This way I can quickly assess the class and possibly spend more time on weak areas such as weapon handling and movement. Each class will be different in makeup and talent and I must judge and treat each one accordingly. Failure to review and refresh the basics can have dangerous, if not catastrophic, results.

## LENGTH

Several factors need to be considered when structuring a class. In the civilian world, many times cost and time away from work are factors that influence the length of the class. Also, generally, police officers cannot get more than five days off before their absence effects their home department, shift coverage, etc. The military has the time to do without a particular soldier as their counterparts are not on a set clock or controlled by a union.

In the end, the amount of time it takes to safely execute the training should be considered. Next, how much can the student effectively absorb and retain should be weighed. These two factors should be primary considerations when structuring a class. In the civilian world, I may have to break my classes into different levels to get all the information across. Sadly, too many times I have witnessed instructors making "levels" for profit purposes and not to ensure adequate or satisfactory training. If I have students perform a drill once, I familiarize them with it. Twice, I am starting to ingrain it. More than two times, I am making them competent to perform the required act.

Further, with multi-week classes, students are allowed the weekends to catch their breath and recharge their minds and bodies for upcoming blocks of instruction. In long classes, it is important to review the prior week's information to ensure it was absorbed before proceeding to a new block. Students should be encouraged to assemble study groups after hours for those who are having a problem keeping up with the information flow. These can be instructor led at first, then student led once a proper structure is established and student teachers are trained and attained. In week three, students can use dry practice drills learned in week one and two to maintain, or improve, their skill set. This applies to firearm handling as well.

### *Length of Day*

The length of a training day can vary due to the intensity of the classes. I have executed physical half-day events that have taxed my mind and body to the point where any further training would have been wasted.

Generally, well-structured eight-hour days are sufficient to get all the learning objectives across to the students and not leave them brain-dead. I have found that a student can intently focus for the maximum of one and a half hours before they need a mental break or diversion. As for structure, I try to do my classroom time first in the morning and then spend the rest of the day doing fieldwork or practical exercises. In the end, if students do not know when to take a break or rest, tell them.

Weather may be a critical factor in this equation. Where I live, summers are hot and humid. For summer shooting classes, I start the class at 6:00 a.m. when there is light enough to see and shoot. I usually stop about 2:00 p.m. because of the heat. From 2:00 p.m. to 5:00 p.m. the wind stops blowing and the range becomes a heat pit of sorts. Students no longer learn, but just try and survive. Others melt down mentally and shut down physically. Instructors need to look at the environmental and physical factors and make smart choices.

In the larger picture, I leave the South during the summer and go North to avoid the heat.

I put more shooting classes on in winter where the temperatures are mild and are easily endured.

### *Classroom Time and the 20:80 Ratio*

As mentioned, I try to conduct my classroom work first thing in the morning, and then move to practical exercises as early as two hours after the class begins. I try to take breaks every forty five to sixty minutes and watch the students' attention span for signs of a needed break. Blind stares, fidgeting, squirming in the seats are a sure sign that students need a break.

If I am running a hands-on class, I follow the 20:80 rule. That is 20 percent classroom time to 80 percent hands on-time. I find this to be a successful equation, especially for type A individuals. The classroom is just enough to prevent boredom and the hands-on keeps the students moving and learning.

### *Field Time*

Setting up smooth and structured field phase session is critical to keep the momentum going and to keep students learning.

Too much downtime and not enough movement and practice will degrade the training and what is absorbed. I run single-station training or multiple-station training. Single should support the entire class. Multiple will handle smaller teams. The multiple-station approach requires more trainers, classrooms, venues, and teaching aids.

## INSTRUCTOR TO STUDENT RATIO

The instructor's skills and the caliber of a student's experience must be kept to a safe and positive ratio. Lecture requires one instructor, and to keep the students on a positive edge, I rotate lecturers from one to several in a given day.

High-risk training may require several instructors. Usually, I will take more students in a first-time law enforcement firearms class than in a civilian class. Law enforcement personnel have been through an academy and have been handling firearms on a daily basis. Officers can implement the proper safety techniques in short order as they have the muscle memory of handling firearms.

Not so with the first-time civilian classes. Most civilians with no experience have to learn muscle memory during the class. Some have learned bad muscle memory over the years and have to "unlearn" poor or unsafe habits. These classes proceed a bit slower with a higher instructor to student ratio for safety. I encounter more malfunctions and weapon issues as the shooters are inexperienced with their weapons systems.

## TESTING

When to test/evaluate? Testing should be performed as a pre-class diagnostic or only after thorough teaching and practice. Many times teachers like to use tests to begin classes or fill class time or to show students how much students do not know. This applies to both written classroom tests and hands-on tests, such as shooting.

If an instructor is going to show students a new system of shooting, a pre-test is useless. The instructor is wasting time that could be productively used in teaching the students a new system.

Tests can be used as a qualification to ensure that students have grasped the material mentally or can demonstrate their skill in performing the tasks at hand. For hands-on testing, remember that it may take several thousand times for a student to gain muscle memory of the drill, but they can demonstrate the ability to perform it after a few hundred repetitions. They will not be as smooth and polished as they could be, but they will be safe in the execution of the drill.

## REFINING THE TRAINING

How to refine training? Many times I can explain to a student what they are doing wrong, but their mind cannot picture or see it. To expedite the learning curve, I use video. As I mentioned earlier, I tape students at the end of day 1 during a tactical handgun course. They have shot for about eight hours and are mentally and physically tired. Pulling a video camera out gives the student an extra burst of energy along with generating a certain amount of stress. With the stress and energy, I get a snapshot of the student's true understanding and ability to perform the drill.

I video the students performing five one-shot drills from the high-ready five one-shot drills from the holster and five five-in-the body, one-in-the-head drills (5/1). With each drill, I am looking for an entire skill set. During the one-shot drills, I look for the entire one-shot sequence and a good follow-through and cover. I look for the smoothness in the draw and acquiring a weapon when the student performs the holster drill. During the 5/1, I look for the student's recoil management of the weapon and stance to see if the student is pushed out of position by the weapon. I watch the body and see if the student starts in a neutral stance or is aggressive. Starting neutral will cause the student to be pushed backward with each shot, requiring longer time to acquire a new sight picture.

I often review the video the next morning when students are fresh and well rested. I ask students to pick out two or three things that they see, to work on today to fix or polish up. After the video review, the class immediately goes to the range for remedial training to allow students to work on their deficiencies.

For each class, I keep a record of the shooting drills or standards. I circle drills that students fail. When shooting the standards, I look at the column with the most circles and those are the first drills I work on as a group. Generally, everyone has the same problems with the same drills. I work those out first, and then start working on individual problem areas by assigning student instructors who are strong in those areas to help the weak students. Sometimes, student instructors have a way of talking to their peers that is non-threatening and is easy to understand.

At the instructor/instruction level of refinement, I use video, formal and informal critiques. An instructor can take video of his lectures to find critique and weak points. Look at your mannerisms, verbiage, dress, posture, etc. Fixing two points at a time with your teaching system can easily provide a more polished presentation.

The same applies to critiques. During the first year of a class, I generally hand out critiques to the students to get their view and feedback. I try and hand out the critiques on day 1 or midway in the course. I want students to keep a daily log or journal of what they want fixed. If I hand out critiques on the final day and the final moment of a class, students will rush through and do a poor job as they have going home on their minds. Also, some students use critiques as a grievance form instead of a tool to shape the class.

I look at critiques before students leave and address any negative issues. In my classes, critique blocks are rated from 1 to 10, one being the lowest. I tell students that if they give me anything less than a 7, I want a short written solution on how to fix it. This keeps them from throwing darts and personal attacks. I use critique forms for the first year of a class and then as needed after that.

## CREATING A POSITIVE LEARNING ENVIRONMENT

Learn how to talk to people in a neutral or positive fashion. The fastest way to shut a mind is to talk down to a student in or in front of a class. This will create unnecessary anxiety in the class/student. I use words such as "polish" and "tune-up" to describe drills they need to work on. Saying, "You suck," does nothing but add a stress layer that prohibits positive learning by focusing the student's mind on his failure of the drill, how it relates to his own self-worth, or what to fix. Students with low self-esteem may take it harder, and it will take them longer to focus when they try and correct their problems. Plus, it is a personal attack versus constructive building behavior.

## STRESS IS WHAT YOU PERCEIVE IT TO BE

As mentioned earlier, I use a timer to induce stress during shooting scenarios. Students perceive the timer as stressful because of the beep. It is almost an annoyance at first. I push students to practice with timers at home and to break their fear of them. Once they become comfortable with the timers, students will go out and buy their own and rarely train without them.

Physical stress, such as exercise, can be used to induce stress for shooting and other tactical drills. Short and long runs and walks at varied paces will induce stress. That, coupled with timers, will cause stress. Putting a student in full tactical gear will cause stress if the student is not used to wearing it.

## KEY POINTS

- Ensure you properly sequence a class for the best and most efficient flow of information.
- Understand your students and be patient with them. Realize they came to you with varied backgrounds, experiences, and motivations. Your job is to funnel information into a positive learning experience.
- Structure your class for the common student.
- Understand and be aware of attention spans.

# 16

# DEVELOPING A SPECIFIC COURSE CURRICULUM

*A logical flow of information is necessary to efficiently fill the mind.*

- **ADVANCED HOSTAGE RESCUE AS A MODEL**
- **A NOTE ON OVERALL INSTRUCTOR TO STUDENT RATIO**
- **CLASS LEADERSHIP**
- **STACKING THE DECK—LEADERS AND INSTRUCTORS**
- **POWERPOINT AND VIDEO TIPS**
- **SUBSTANCE VERSUS FLUFF**

## ADVANCED HOSTAGE RESCUE AS A MODEL

Now that you have a general understanding of the guidelines that shape a tactical class, I want to offer an example of how to adapt those principles to a specific course with a larger scope.

No matter what topic you are covering, remember that the more logical and easier the information flows, the more the student will absorb and assimilate.

My core course that I have probably taught more than any others is Advanced Hostage Rescue. Without giving away tactics, I will use the course as a model to describe how I structured the course maximum flow and stair-stepped learning.

---

*DAY 1 INTRO/ADMIN/HR OVERVIEW/EXTERIOR MOVT 12/07*

| | |
|---|---|
| • My Intro | |
| • Group Introduction | |
| • Roster | |
| • Team Breakdown | |
| • Team Photos | |
| • Safety—Paintball/Vehicles | POWERPOINT |
| • HR Cycle | POWERPOINT/COBB COUNTY |
| • Combat Mind-set | POWERPOINT |
| • Equipment | |
| • Weapon Safety Inspections | |
| • Stack-Knee | |
| • Movement/Formations (Flare) | VIDEO/POWERPOINT |
| • Actions on Contact | |
| • Suspect Manipulation | |
| • Officer Down | |
| • Scenarios | VIDEO |

*DAY 2: CQB*

| | |
|---|---|
| • Review Day 1 Video | |
| • Safety | |
| • CQB | POWERPOINT/VIDEO |
| • Flash Bangs | POWERPOINT/VIDEO |

- Assault Breaching POWERPOINT/VIDEO
- Scanning Drill VIDEO
- Flash Bang Drill
- Hostage/Suspect Manipulation Exercises
- Closing Down a Room
- Force on Force
- Medical Emergencies (Team Demo)

*DAY 3: INTERIOR MOVEMENT*

- Review Video Day 2
- Safety
- Interior Movement POWERPOINT/VIDEO
- Corner Clear
- T Intersections
- Hallways—Single
- Hallways—Double
- T intersections
- Breaching Setup
- Stairs/Elevators/Restrooms
- Officer Down
- Video Scenarios

*DAY 4: TEAM CQB/PLANNING*

- Review Video Day 3 2+ Hours
- Safety
- Assault Planning POWERPOINT/VIDEO
- Team Leader Duties
- ATL Duties
- Sniper Responsibilities
- Individual Responsibilities
- Stealth versus Dynamic
- Known versus Unknown Location

- Multibreach Linkup and Coordination Points
- Anti-Fratricide Teaching Points
- Hostage Evacuation
- Medical Emergencies
- IEDs
- Assault Breaching Video
- Assaults
- Team Planning PE

*DAY 5: SCENARIOS*

- Deliberate
- Deliberate with Compromise
- Stealth
- Deliberate with IED
- Emergency Assault X 4

---

I developed this class in 2001. I wanted a comprehensive class that was structured to the worst-case scenario that could be used as a template plan for training and for use in other missions such as High-Risk Warrant, Barricaded Person, Search Warrants, etc.

Developing a course is like stacking bricks in the corner of a building. If you leave an unorganized pile, the bricks are there, but they take up a great deal of space and cannot be accessed efficiently. By stacking them in a logical order, they take up less space and can be retrieved much more easily. This is the same as class information. If the instructor takes the time to logically put the information together, it will save the instructor time and the student will be able to store more information.

As I noted earlier, I commonly ask students, what is the difference between taking gunfire on a hostage rescue, high-risk warrant, and a barricaded person mission? There are no differences. Getting shot at is getting shot at and tactics and techniques must be the same for each mission. The only difference is when I push through the threat in one mission and the other I isolate and contain the threat. I, then, develop my tactics and techniques to deal with all the possible scenarios.

## *DAY 1*

Introductions, Safety, Hostage Rescue Operations, Combat Mind-set, and External Movement are the main topics on day 1. I introduce myself and give a brief background of why I am qualified to teach this course. I give my goals and objectives. In my Advanced Hostage Rescue, I advise the class that I am trying to teach overall team and organizational concepts for 75 percent of the class and 25 percent will be individual techniques.

Safety is the first class I teach, and it is my intention to set the stage for this class and future operations. I give the student my basic set of safety rules, then give a case or show video of the rules being violated with the graphic results. I also tell the students why the safety rules prevent accidents. I provide safety protocols for the class that they can take back to the student's agency.

Hostage Rescue Operations is my next block of instruction. I show the "entire elephant" about hostage rescue operations. I discuss hasty and deliberate hostage rescue. Next, I discuss the overview from start to finish. I show a video clip every few slides to keep their minds from becoming bored. Even the most attentive student can lose focus if I do not "bump" their mind every few minutes.

Combat Mindset is my next class in the course, and I set the stage on how I want them to think, train, and react to violent confrontations. It is probably one of the most critical components to the course and may help save their lives in high-risk situations.

After a lunch break, I come back to the classroom for the External Movement Class. Knowing SWAT and the type A personalities that compose these teams, I cannot keep them in a classroom more than a day. I give a brief lecture on how to move from a drop-off point, or from a vehicle to a breach point, and how to handle the various problems that they may encounter. I teach five simple scenarios to respond to and instructions on how to handle the scenarios. The class moves to a training area, performs our safety checks, then walks through the drill. I tell the TLs to take their teams to four different training areas and rehearse the team actions on contact outside their breach point. This gives the TLs time and a pride in ownership when working with their teams. There is incentive to perform well and practice since executing the drills is filmed and the entire class will review it the next morning for review.

I create five common scenarios and give the solutions to the students to practice. Tactical teams routinely move from vehicles, or on foot, to breach points (doors or gates) and need a plan to deal with common problems. By forcing students to develop these plans, teams are beginning to get ahead of the game. These techniques can also be used on other tactical missions and provide a system that can handle various common tactical problems.

## *DAY 2*

Day 2 consists of review, Close Quarter Battle and Flash Bangs. I begin the class with a video review of the drills from the previous day. I look at the video in a mechanical fashion in an effort to determine problems with fields of fire, exposure, muzzle sweeping, etc. I want students to learn from the video and not cringe when it is played. Again, I ask, "How hard do you want me to critique?" The answer is usually, "Hard." As an instructor, I must remember to temper my responses and choose my words and vocal tones to keep their minds open. I ask them to give me some points about the video clip to see if we are seeing the same things.

When correcting items, I ask students to provide options on how to better solve the tactical problem they are facing. I ask them what the priority of the drill is at this time. I want them, in their minds, to establish a simple one, two, three priority of sorts to solve tactical problems. I want them to think and offer solutions. Once a point is made, I try not to hammer it if it is done incorrectly again. At the conclusion of all the teams' reviews, I ask the class to list on a dry-erase board five points they would polish drills if they were to perform them again. Time permitting, the class rehearses and practices the drills again to fix the issues.

The next class I present is CQB, which deals with how individuals, two-person elements, and teams enter and dominate rooms. Over the years, many CQB systems have been developed and evolved. I show students the core systems and discuss why I do not use the others, either for safety or for lack of mutual support in the room (looking at individual sectors instead of the entire room). I show several choreographed drills on video of the drills being performed to standard. The final portion of the CQB class deals with handling people and medical treatment.

Flash Bang instruction covers exactly what its name suggests. Dealing with flash bangs is the final class of the day, and I give a short but comprehensive class on diversionary devices that can be used in conjunction with CQB. I teach a simple system, show the students funny bloopers of individuals in past classes, and encourage them to put more effort in practice/rehearsals today. All the diversionary devices used in the class are dummies as deployment is a tactical skill that would take a half day on its own to teach.

After lunch, students report to the training site for practical exercises. I begin by having two-person teams enter a room filled with paper targets to see the one with the guns. I visually overwhelm the individuals and show them how to break down a room into a simple scanning sequence. (I video this portion for review the next morning.) The video shows me entry speed, head-eye-weapon relationship, and their scanning sequence. Next, I break teams into two groups. I take one group consisting of two teams in a force-on-force scenario where one team plays the people in a room and the other team enters and manipulates the other students. Reality is that 99 percent of the time, police are putting their hands on people and not shooting them. Students must know how to do this efficiently.

I demonstrate how to enter around people, how to move, use cover, etc. I have the two teams switch allowing the other team to gain hands-on practice. Simultaneously, the other TLs are practicing two person CQB, team CQB, flash bangs, and the discrimination room in an attempt to refine their new team's SOPs. I want students to know how to run rehearsals and to take that skill back to their teams.

After both teams have performed the hands-on drills, I bring all the teams together and explain how to tactically handle a medical emergency. I introduce bandages, tourniquets, medical scissors, etc. I show students how to "layer" their gear in the proper sequence for use. I stress that if they cannot get to the gear themselves, without help, it is useless. Students are taught to prepare their wound kits for quick deployment. This consists of unwrapping and preparing bandages with knots in the end for use with a gloved hand.

I demonstrate a sequence for establishing security, exposing the wound, direct pressure, bandages, and movement of the patient. I bring in a student wearing old pants and have one team go through the scenario where I squirt blood all over the patient and the hands of the treating officer. The class works through the procedures for processing a medical casualty and transporting a patient off the scene. I constantly remind students that they are the first responder for patients, including themselves, their buddies, innocents, and the suspect. I want the students to prepare for and expect worst-case scenarios.

## *DAY 3*

Day 3 begins with a video review and Interior Movement class. I usually start the video review with a sound check of my computer and speakers by showing a humorous video to establish an open mind-set and to relieve any anxiety that students may have about watching themselves on the big screen. After a short clip, I review the two-person room entry videos and move on to the class.

In reference to structure, I teach students how to move to the building or the structure on day 1 from foot or vehicle. Day 2 involves taking down rooms, diversionary devices, and handling people in rooms. Day 3 involves moving down hallways and "processing" rooms and controlling people in them. Most actions on this day will focus on how to handle situations in a hallway in an efficient manner. Similar to day 1 and exterior movement, students will be taught how to tactically move around a ninety-degree corner and handle situations encountered in the hallway.

Again, I use a PowerPoint program with video clips to show how I wish the drills to be performed. Then the class moves to the training site and conduct safety checks. I perform a reduced speed demonstration before sending teams to their own areas to practice. I give students two hours to work with their teams and role players before I begin filming them. I walk around and answer any questions that may arise. At a designated time, I gather role players, paint-marking gear, and begin filming five different scenarios with one team at a time moving down the hallway. The other teams are not allowed to watch and must wait their turn to execute the drills. Role players are briefed and told where and what to do along with being reminded not to alter the scenario. I have even gone to the point of putting a piece of tape on the floor to mark where an action is to consistently happen.

The reason I put hallways on day 3 as opposed to day 2 is simple. I want students to practice entering and dominating a room every time they enter a room from a hallway. This builds "mental" muscle memory and can reinforce the prior days' training with every movement and room entry.

My goal is to teach a team to quickly and efficiently handle a problem and move on. After each of the teams has gone through the scenario, I put two teams together and have them move as a unit and solve problems. I film from the front and the rear to catch all the action. After both groups have completed the exercise, I conduct a debrief of the day and release the students.

### *DAY 4*

Day 4 is "a catch your breath" day where the class spends the entire day in the classroom and prepares for day 5. I begin day 4 by reviewing the video for the first hour, from the rear for the second hour. Students can see what happened, how much time it took to solve the problem, how much the team exposed, etc. By watching the video from the rear, the class reinforces what happened and gives suggestions on how to fix it.

I begin the instructional phase of day 4 with an Assault Process class followed by an Assault class. The students have worked the entire week as individual teams and now it is time to put the package together. I talk about o*perations orders* (OPs), why OPs are important, and who does what according to the OP order. I review how to plan and execute multiteam raids and the safety factors involved. I have TLs start by using a dry-erase board, sketching a building with a simple floor plan that has no windows to determine team flow. I have the ATLs plan using a building with windows. Complex targets are also discussed.

Using this technique, I lay the groundwork for the planning I want done on day 5 for the scenarios. The students are already practicing a system of planning and briefing OP orders. I finish the class and discuss the responsibilities of everyone on the assault, who does what in both planning and during the takedown of the target. I give sample briefing formats and communications procedures.

The last part of the day is planning and briefing target folders of targets I have sketched and photographed. The targets ranged from professional businesses, to restaurants, to common venues where past incidents have taken place. Each team has a different folder and plans accordingly. This exercise builds planning and briefing skills and forces them to figure out how many team members and what equipment they need to actually perform the mission. During this phase, each team briefs a hasty and deliberate plan to the other teams and they are allowed to ask questions. By using four different targets where incidents have taken place in the past, teams get a template plan of sorts on how to takedown each target. This builds for success for the day 5 scenarios.

### *DAY 5*

Day 5, the class hits the ground running. Teams show up to a target early and I allow the leadership one to two minutes on the target to sketch it. Students begin planning for deliberate assaults. When their plan is complete, they brief their teams and perform rehearsals if time permits. Once that is accomplished, I give them permission to hit the target. I walk with the designated commander and use him as a second set of eyes. I start with all teams assaulting the target and practicing how to control, pass information, and consolidate personnel during the scenario. The first scenario is repeated because many students fail to grasp how to put their team and individual skills together on the first hit. The second hit usually goes much smoother.

After each hit, I conduct an AAR with everyone involved. Each TL briefs what the team did and why. I go around the room, not looking for who did what, but rather what worked and what did not and how to fix it. I suggest that if they fix two to three items per scenario, by the end of the day, they have made great strides in their tactical drills.

After the combined hits, the class performs individual team hits with the follow-on force coming in one minute after a REACT Team conducts an emergency entry to neutralize a threat. By executing the full-blown assaults first, teams learn what to do and how to do it. They also know what is relevant and what is not.

## A NOTE ON OVERALL INSTRUCTOR TO STUDENT RATIO

I developed the course to be taught by one instructor. Some say a 1:24 instructor-to-student ratio is too large. I disagree and believe this depends on how well the information is organized, how it flows, and how the class is structured. Additional instructors can be added for live-fire operations.

This course is a paint-marking course only. No live-fire training is performed. Safety checks are critical. I show the class how to perform these checks using their internal chain of command in a way that should be replicated at their home department. I push the responsibility of safety back on the individual student and team, where it belongs. I require the individual to check his own weapon and equipment, those of a buddy or ATL, and finally their TLs. I serve as the final check for the safety of the class.

Early on in the class, I establish protocols as to the consequences if I find live ammunition during a paint marker training exercise. This would be the immediate release from the course of all three individuals in the safety chain. Students in the past have commented on a need for a "safety officer." My question to them is, "Do you use one back home during real operations?" The answer is no. My point is that I teach a system that they can take back home and that works on the street.

I encourage questions at any time and answer all with respect to the student, no matter how simple it may be. Embarrassing a student with a flippant response is unacceptable. It creates a negative environment that does not promote dialogue.

## CLASS LEADERSHIP

One of the biggest voids I have seen in any training, including law enforcement, military, and civilian organizations, is the lack of leadership and how it should interact with the class being taught. There are few leadership classes out in the world and only a few being taught by qualified leaders. I believe that leadership should be taught at all levels to help fill this void.

## STACKING THE DECK—LEADERS AND INSTRUCTORS

To ease my workload and to help distribute knowledge across the class, I appoint TLs who are strong, who have seen my system before, or who have taken my classes in the past. These students understand my work ethic and what I demand as far as safety, rehearsals, etc. I keep my training and leadership structures in all my classes simple and streamlined, because simple is best.

Picking strong TLs allows me to disseminate information to only four individuals. TLs are responsible for checking their teams for accountability and equipment. Instead of trying to keep track of twenty-four, I only need to keep track of four individuals. If a TL is late, the ATL is responsible for leading the team. This is a simple and effective system that mirrors real-world operations. For some students, this is standard practice, for some it is an eye-opening experience.

## POWERPOINT AND VIDEO TIPS

I try and show a video every five to six slides during my PowerPoint presentation to keep students' minds open and engaged. I use video slides of students doing the drills correctly and of students or incidents that resulted in mission failure, death, or injury.

## SUBSTANCE VERSUS FLUFF

I've mentioned this earlier, but it bears repeating. Excessive use of "war stories" is a sure sign that an instructor does not have enough material to present or has not thought out a class. While war stories are important, relying on stories to entertain students leaves them wondering what they have learned. War stories should be used to back up and illustrate points, not to kill massive amounts of time in a class.

## KEY POINTS

- Look at the "whole elephant" first.
- Structure your next class so each session builds on and complements the next.

# 17

# VALIDATION AND REFINING YOUR INSTRUCTION

*Honest AARs and positive corrective training is the only way to induce excellence.*

- **SECOND SET OF EYES**
- **PAINT-MARKING GUNS**
- **ROLE PLAYERS**
- **VIDEO RECORDINGS**
- **DRILLS**
- **TESTS: WE MUST TRAIN BEFORE WE TEST**
- **RECORDING POSITIVE AND NEGATIVE**
- **CRITIQUES**
- **DRY-RUN CLASS**
- **VALIDATING TRAINING AND PRACTICAL EXERCISES**

## SECOND SET OF EYES

In my tactical classes, I create a "second" set of eyes during every course. My goal is to teach others how to see what I see and what to look for during training modules. I believe in creating and grooming students into potential instructors and expanding their awareness. I create these additional eyes by appointing team leaders to help monitor their teams during training, rehearsals, and exercises. I will also elevate team leaders to the next position of authority, and that is the tactical commander. I believe that leaders should know all the positions below them and one above.

As the tactical commander, students are put in the uncomfortable position of relying on their TLs to be aggressive, to take down the target, and to get critical information back to the tactical commander. When that information does not get communicated, I hold TLs and the tactical commander accountable. Students posing as the tactical commander quickly realize how important the job of TL is, and when I put them back in that position, they understand that it is important to be aggressive, decisive, and accomplish the mission in a rapid and efficient manner.

As mentioned earlier, I also use the tactical commander during AARs. My thought on this is that students may sometimes feel disheartened that I am continually critiquing them. I want them to hear the same message from one of their own peers. Sometimes, a different voice with the same message will be heard.

## PAINT-MARKING GUNS

The use of paint-marking guns has taken force-on-force training to new levels. Besides being used in areas that cannot handle live-fire ammunition, paint-marking guns can be taken anywhere to ensure training is accomplished on real-world targets.

The additional benefit is force-on-force training can push the student to higher mental and physical levels of performance. Mentally, students are taught in their career to never point a gun at a person during an officer's entire career, then we expect them to do great things on the street during high-risk real-world situations. This type of training is unrealistic at best and sometimes counterproductive. With paint-marking guns, we can allow officers with "mentally real" weapons to approach, train the weapon on, engage, and terminate threats through controlled role play.

This mental training is as critical as the physical act. They both go hand and hand. Instructors must train officers to physically control all their emotions and stresses during this time, deliver a critical shot when required, and put many pieces to the puzzle on demand. Paintball weapons can be used to measure safety and accuracy of the officers. These weapons will show whether or not the officer's academy, range-fire, and in-service training are building their skills to a satisfactory level or hampering their performance.

## ROLE PLAYERS

Role players are critical to bringing students to the next level of training. Role players add the next dimension of interactive targets and are as good or bad as their controller. The controller can either be the trainer, a designated "head role player" or someone standing outside the exercise running a group of role players.

### *Role Player Controllers*

I use an experienced role player to control other role players when I have two or more role play actors. I pick either someone with experience or one that has the most rank and leadership ability. First, I put them in charge of safety and have all role players checked for weapons. Next, I go over the general scenario and break it down into subtasks. For example, my scenario might be a hostage rescue situation where the location of the shooter is unknown, and this might be the block or wing of a school. I have already looked at the area and marked out-of-bounds areas with yellow "do not cross" tape. This is done for safety and the skill level of the students. They may not be ready for a large and elaborate area.

After safety checks are done, I give general guidance about what I want to happen during the scenario and the intensity with which I want it to happen. The first run of the day is usually the hardest for the team since teams have to put together the entire package. I try to select a simple exercise where all the modules come together, such as breach points, hallway movement, CQB, reconsolidation, etc. I tell role players, I need a single shooter, a medical, a runner, etc. I then ask who wants to do what. Some role players make better victims, with moulage and blood, and act out a part. Some role players are aggressive and like to be a shooter. Others prefer to be a victim or wounded hostage. I may not give them a specific task during the scenario other than being an unconscious body. In the case of the running hostage, I advise the role player to initiate their action when you hear them coming down this hall. I want you to run toward them and comply with what orders they issue you.

Reference medical scenarios, I have them lying in a specific spot, moaning, trying to get the teams to drop security and lose focus of the real mission. For role players who are shooters, I tell them to use common sense and explain what that is. I tell them that if they get into a mop closet and try and shoot it out with the officers, it will hurt. Find areas where you can engage from a distance and either die in place, retreat into a room and fight it out, or surrender. Also, I instruct role players to slide the weapon away when you surrender/die unless this action is part of the scenario. If they fail to do this, the officer may shoot you at close range.

For large groups over fifteen people, I assign multiple controllers. If I have an outside action and an inside action, I have two controllers overseeing the different areas. Larger role player groups have to rehearse the actions to ensure they are getting the desired effect. For control purposes, I issue radios to role players to keep in contact and maintain timelines so that they can initiate certain actions I want to happen at a given moment.

Be wary when dealing with individual role players and their actions. Some "professional" role players are those who do it often for fun and try to "get" their buddies. These are the ones that you have to watch.

My goal as instructor is to bring positive training to my students and not to have them harassed. I have put role players down myself for doing silly actions not in the script, spontaneously making them a casualty by squirting blood over them from a bottle I kept in my pocket. Horseplaying role players can disrupt the scenarios and learning points.

I cannot add chaos to a script or scenario unless the students performing the exercise get the basics down. As students improve, I add more for students to sort out. I start with medical emergencies, add runners, officers down, IEDs, etc., to make the scenario more encompassing. This is done at my pace and not that of the role players. If I were running a singleteam scenario, I would want the same action or acting for each team. Consistency is the key. It does not matter if the role players get bored. It matters that they give consistent performances for each team, especially if I am videoing the scenarios. It will become readily evident, during the video review if the scenarios change drastically. The role player's job is to assist in training and not to do their own thing.

## VIDEO RECORDINGS

"Video does not lie" is one of the truest statements I have heard, and one I cannot emphasize enough. If used properly, video is an incredible tool. I use video extensively on the flat pistol range, during Advanced Hostage Rescue training and Tactical Team Leader courses. After firing on the pistol range, I can show details to a student during a classroom video session to illustrate all the little things their mind cannot absorb when the gun is going off. I tape them from the side to show students how their body is absorbing and managing the recoil of the weapon. I zoom in on their hands to examine their grip and see if their arms are locked. Using video in this manner is the fastest way to get points across to a student.

During classes for Tactical Team Leader, I record the loading, firing, and unloading sequence of the line. I critique safety issues and poor firing positions. I see whether or not students are performing a good follow-through and cover when firing. Finally, I note if students are safely unloading their weapons and if the TLs are checking their team members.

In my Advanced Hostage Rescue class, I run a day of hallway training, which is the most difficult to learn and grasp. I video from the front and from the rear to show the student's problem areas. During these paint marker drills, students may have different jobs and a five-person team may only have one to three people actually seeing the entire action and what happened because some are in the room. One person may be turned around providing cover to the rear; another may be entering a room when the hallway action takes place. By reviewing the video the next day, all students can see what happened during the scenario and how much time it took to accomplish certain actions. Students can make adjustments accordingly. I also show them how they and the scenario appeared from the rear. Also, by showing this video the next day, students are fresh and rested. It is an incredibly powerful teaching tool.

## DRILLS

Battle drills or Immediate Action drills are used to solve problems that students may face during scenario-based training. I break scenarios down into easily understood and practiced modules where students can solve a problem and move on. It is critical in high-risk situations to quickly and efficiently solve a problem and move to limit exposure time.

The key to developing these drills is to keep them simple, multipurpose, and common sense. These drills should be able to solve tactical problems during different missions which officers encounter. For example, hallways are the same in Hostage Rescue, High-Risk Warrant, Barricaded Person, and Search Warrant. The principles remain the same. I should not have four systems to clear a hallway, at the most two. One system should be dynamic and one should be with shields. Three out of the four scenarios can be cleared dynamically, but one may require the use of shields. This narrows the hallway clearing down to two simple systems. As a note, the dynamic methods can also be used to teach patrol officers active shooter tactics. I have a system that works. Why reinvent the wheel?

Remember, there are many modules or components to a tactical mission. Keep drills simple and efficient and realistic. Develop tactical drills for worst-case scenarios and validate with role players, video, and a stopwatch. Evaluate how much time it takes to execute the drills and how much time officers are exposed in the hallway. Remember, video does not lie.

## TESTS: WE MUST TRAIN BEFORE WE TEST

Too many times I have witnessed instructors testing before they have adequately taught the material and have given students time to practice and become semiproficient. Many times this testing is done to "show" the student how much an instructor knows. It is simply counterproductive and retards the student's learning curve. Testing before training consumes valuable training time that the instructor will never get back (see chapters 3 and 15 for more details).

Further, the word test generally creates stress and anxiety in any student. Tests can be written or hands-on. During tactical training, I am reluctant to have extensive written tests other than to create a paper trail for the bean counters. What is critical in our profession is not that a student can memorize the big four safety rules, but that the student can put these rules into practical application around fellow officers and citizens on the street. Many times tests are given in writing when they should be accomplished with a hands-on scenario. You find this with instructors who are more comfortable teaching in the classroom rather than in the field.

My emphasis on weapons safety is not demonstrated by how well a student scores when writing down the safety rules, but how he responds to a simulated high-risk scenario with his weapon, how he handles the scenario, how he moves around fellow officers/civilians, and how he "safeties" and reholsters his weapon. A written test in this case is a waste of time because the street has a much harsher way of finding out whether a skill set is there or not.

## RECORDING POSITIVE AND NEGATIVE

Usually, instructors record only the negative during a student's performance and not the positive points. I believe that both need to be addressed to get a well-rounded perspective on the individual. Praise good performance. Tactfully correct issues. Ensure students understand the same problems that the instructor sees and understand how to fix them (for more information, see chapter 9).

## CRITIQUES

Proper critiquing of students can make or break the instruction. As mentioned earlier, critiques should be constructive and nonpersonal when criticizing and personal when praising. Personally attacking a student during a critique will shut down the student's mind and put him on the defensive. Ask a question, such as, "What problems did you encounter?" This keeps it neutral and also allows the student to voice any concerns that the instructor needs to know. Allowing the student to speak ensures that the instructor and student are on the same wavelength.

When critiquing students, I try not specify more than two or three items to "polish up." Again, terminology can make a big difference when talking to students. "Polish up" is another phrase I use to keep the emotions down and to keep it neutral. If I were to say, "What are you going to do to unfuck yourself on the next run?" I would send the student into a defensive and unproductive mind-set. The student is now thinking:

1. He thinks I am stupid.
2. Oh no, another run and chance to screwup again.
3. If I screwup again, I will lose my mind.

Part of the instructor's job is to put students at ease mentally so they can focus on the task at hand. There is enough natural stress in a class, items such as weapons safety, movement and shooting with weapons around peers, discrimination, hitting the target, etc., without pouring more on.

## DRY-RUN CLASSES

Dry-runs cost you nothing and do everything:

- They build muscle and mental memory as an individual.
- They build muscle and mental memory as a team.
- They allow instructors additional time to inspect safety and target setup.
- They promote good habits and work ethic among the force.

In my early days of Special Operations, I might deploy or live under close conditions for months. The army had limited live-fire ranges and sometimes did not have spare ammo to fire. I dry practiced for twenty to thirty minutes a day to keep mentally and physically sharp. It worked. For shooting, the only thing that cannot be replicated is recoil management. Having said that, I was almost guaranteed a first-round hit when the time came to do business. I remember times where I was deployed and the only thing I could do was dry-fire to prepare for unit shooting matches and came back and won in-house shooting competitions upon my return.

For team movements, it keeps the teams working off each other and flowing well when the time comes to take down a target. I remember many times doing one to three hours of dry runs at a time.

## VALIDATING TRAINING AND PRACTICAL EXERCISES

Training that includes hands-on tests must be validated prior to administering to a student. I put my own cadre through the class and see how they perform to get a baseline for the new students. If the event is time driven, you might add 20 percent more time to that of the cadre time. The same applies to marksmanship or shooting evaluations. Once the first group of students goes through the test, the instructor must also look at their scores to see if the times or standards applied fit the target group.

In today's society, where academies sometimes water down standards to pass more recruits to the street, an instructor must effectively and efficiently bridge the gap and bring students up to a higher operational level. An instructor must understand this as an evaluator and realize that students may not be able to make the mental, physical, or tactical leap during one of your evaluations and must be retrained to standard.

## KEY POINTS

- Buy two cameras and a projector to show videos on the wall (or screen, if available). Video recording is an invaluable tool.
- Structure classes for the common student and validate tests with cadre members. Be sure to include physical, technical, and written elements. Adjust standards in accordance with the level of the students' training.
- Understand the attention span of students in high-risk courses.
- Learn how to critique students in a positive and nonthreatening way.

# 18

# MANAGING HIGH-RISK TRAINING: GENERAL CONCEPTS

*One aw-shit will wipe away all the attaboys.*

- **THREE LEVELS OF COMBAT**
- **STAIR-STEPPED TRAINING AND INSTRUCTOR CONTINUITY**
- **DIAGNOSTICS**
- **SAFETY ISSUES**
- **SPECIALTY TEAMS**
- **FLARE AND ONE-MAN AMMO**
- **COACHING, COUNSELING, AND REMEDIAL TRAINING**
- **TEAM AND ORGANIZATIONAL TRAINING TIME MANAGEMENT**
- **EQUIPMENT MANAGEMENT**
- **YEARLY AND REPETITIVE TRAINING**

## THREE LEVELS OF COMBAT

As mentioned in the introduction to this book, the success of any tactical mission depends on the application of proper training at the individual, team, and organizational levels. This same principle applies to the management of high-risk training. It may not be the bad guys shooting at you, but the danger is still there.

At the *individual* level, students should have attended basic, intermediate, and advanced level classes that promoted initiative and self-practice philosophies to ensure that the individual's tactical and technical skills are maintained. Technical skills can range from physical fitness to shooting. Students should be taught how to practice, self-analyze (standards/video), and maintain their skill set (dry-fire/physical training).

At the team level, ATL and current TL should have attended basic, intermediate, and advanced team-oriented training and also leadership courses in how to manage and implement this training and sustain the team's technical and tactical skill set. As mentioned in the previous chapter, Advanced Hostage Rescue is geared 75 percent to the team organization and tactics and 25 percent to individual skills. I also teach a certain amount of command and control.

At the *organizational* level of training, mid- and senior-level leaders need to know what the individuals, teams, and subunit leaders are doing and why. I cannot emphasize this point enough: *Any individual in combat should know their job, one job down and one level up, to be a productive member of a team.*

Too many times leaders are "plugged" into crisis situations without any training with subordinates on roles or how scenarios should be played out. Too many times leaders come in, attempt to take charge, and direct training with their limited knowledge base and pace. This can get officers and civilians hurt and killed.

A simple example is when an officer is shot down on a patrol call and left to bleed for hours, resulting in the death of the officer. Chiefs radio in directions not to do anything until they arrive. The downed officer remains under direct gunfire from a bad guy and is still bleeding. It can take fifteen to twenty minutes for the chief to make it to the scene. Meanwhile, the situation has deteriorated, as has the officer's health. An officer can bleed out in five minutes if the wounds are severe enough, and while this is happening, the bad guy may be establishing a better position to fire on other officers. In a scenario where an individual or TL knew the role above and below them, a proactive team member can make an on-the-spot decision to recover the downed officer and execute the plan because they have already trained for it. Chiefs should already know what the plan (or drill) is and have blessed it long ago.

## STAIR-STEPPED TRAINING AND INSTRUCTOR CONTINUITY

Most of the injuries and accidental shootings I have witnessed over the years have been from disjointed training programs, lack of safety protocol, and simply wanting to go too far, too fast. Disjointed training programs are assembled haphazardly or handed from one instructor to the next without an overlap of instructors or a continuity book to help the new instructors out. I have watched large corporations writing up six-week training programs with a close cell of trainers who would hand off the package to a totally different cadre to implement the program. This disjointed approach induces many problems. For this to work, some of the instructors involved in writing the program should also be part of the training cadre to pass on the concepts of the folks who wrote the material.

The same idea applies to individual instructors. An instructor should overlap with new instructors who are assigned to take over and teach the material. This way questions can be asked and answered and intent can be determined. If this cannot be accomplished, a "continuity book" should be written by the current instructor for future instructors. This continuity book should give the "how to" for new instructors. A simple example is that of a hand combat program. When assigned to Special Operations, I was assigned with another instructor to put together a hand combat program for new students. After we completed the assignment, we made a continuity video tape of all the moves and lessons for the instructor who was going to take over the program. This way the new instructor had not only a written lesson plan and a visual lesson plan to use.

This simple technique helped keep the new instructor on track with a proven program. He could adjust, delete, add, or change any parts to the program, but should provide his replacement with the same continuity program he was provided.

## DIAGNOSTICS

I usually do not perform diagnostics as a general rule. I found that I wasted too much time in the process and that I have to go over the basics in any case. For a new student class, I must start from the beginning. If I am teaching an intermediate or advanced class, I just show the students what I want them to know and use. This philosophy also applies to basic-level classes as I want to start with a clean slate and not see ten different systems being used.

## SAFETY ISSUES

As a rule, I always cover and reinforce safety and the basics in every class. I am in a high-risk profession with students from every type of background. They can come from small or large departments with vast or little experience. I must elevate them all to the same level at one time. Safety is a never-ending process that needs reinforcing all the time. Why? The same mistakes keep happening. I use a standard one-hour safety briefing in all my classes. I have not changed the rules in eight years. I can track any "accident" back to a violation of one of the rules. If I set the tone for the class from day one about safety, it will send a message of importance to the students. Safety is an integral part of training and must be a simple and routine system that allows the instructor to train and focus on what is important. Safety systems allow the individual and team to train as close to the edge as possible. Safety protocols should also be stair-stepped and students should earn their right to move to the next level by performing in a consistent and safe manner at the current level.

Safety can take the form of rules and regulations and also that of devices. One common device for safe weapon handling is that of the chamber blocking device. Used when an officer is no longer required to fire on the firing line, it is inserted into the chamber or magazine well of the weapon to prevent accidental firing. Many are bright colors so other officers and range safety officers can see them. The Chudwin Chamber Blocking Device (CBD) is a professionally made device. Improvised devices can be made from bright yellow nylon ropes and cut to fit.

Why are blocking devices important? The first part of any rifle firearms program is that of basic zeroing, which entails adjusting sights. Students will have varied experience levels. I must bring them up to speed with the same procedures. It would be tragic if a student was injured or killed adjusting his sight on a weapon that was still loaded. By inserting a blocking device, the weapon is incapable of firing when a student is adjusting the front or rear sight. Students may have little or no weapons experience at this stage of training and the instructor must look out for their welfare and teach a system that they can use back at their home agency.

These blocking devices should be used for a day or two until a student earns their right to carry a hot or loaded weapon. I watch students' muzzle and safety awareness during the day and ensure that they reinforce safe handling procedures each time they manipulate the weapon.

Another safety device is a paint-marking weapon (see chapter 17 for more details). It can be used as an intermediate tool to take students from dry-practice to live-fire. Again, students have to earn their right to go live-fire. There is always the one student that just does not get it during training when I wish to move on to more advanced skills. I can issue the problem student a paint-marking weapon while everyone else fires live rounds during the training and exercises. This way the student in question continues to train and, hopefully, improve. If they do not, I can remove them from training and I have not endangered any fellow students. Also, this allows the training to continue at my pace and not be held back by one student.

## SPECIALTY TEAMS

*Specialty teams* are teams that have a primary mission in one area as an additional assigned task. For example, if a unit has several different missions, instead of training everyone to be an expert in vehicle assaults, I would assign one team to be the "experts." When the time comes to train the entire unit in vehicle assaults, this team would be the lead in this specialty skill. This way I can assign multiple teams different missions and be more efficient with the training time. Instead of an entire group performing vehicle assault training, one team trains and brings back the information and replicates the training to other teams. It is cheaper in the long run and the students all gain the information. One team will excel at vehicle assaults, surpassing anyone who is mediocre.

## FLARE AND ONE-MAN AMMO

Managing high-risk training takes experience and attention to detail. Safety must be observed and students must earn their right to ascend to live-fire drills.

I teach the Tactical Flare for engaging individuals on the outside of a structure when moving to a breach point with a tactical team. It is a maneuver that allows everyone to see and shoot. Individuals must move behind the shooter in front of them to their own firing position in relation to the threat. Certain individuals give verbiage, others cover, others go hands-on. I start this drill dry and with a single role player. Once this is mastered, the team proceeds to paint-marking guns and a live role player. Then they are ready for live-fire. This is where things can get lively. A typical team consists of five officers. Five guys shooting live around each other at the same time can be hairy.

A simple solution is to use a single knock-down target and to only give one officer live ammo at a time. With each drill, the officers build muscle memory on how to move and execute the drill. With only one officer shooting, the instructor can easily evaluate and control the students. I determine accuracy by checking the target each time. If this officer does the drill to satisfaction, I give another officer live ammo and check his performance. I continue this process for each person on the team until all have shot and proved their competence. Then, I give everyone live ammo and have them go at half speed.

This simple technique allows me to control and evaluate one shooter at a time and give each the attention and quality checks required. Once the shooters/students have this drill down live fire, I go back to paint markers and role players to start throwing in possible contingencies to the problem.

## COACHING, COUNSELING, AND REMEDIAL TRAINING

I prefer to use team members and TLs to aid in coaching, counseling, and remedial training. As a former TL, I prefer not to take a new guy on and be his mentor. The problem with that was I lost my focus on the rest of the team. Instead, I assigned another team member to work with them. For example, if the new team member was weak in their pistol skills, I assigned them an accomplished shooter. I checked on the assigned trainers from time to time to see how their student was performing.

I prefer to use my internal team resources to help develop and polish out our new member for several reasons. First, it will show that I care and that there is a standard. Second, it will set an example for what to do when they are a leader. I will also give the trainers the authority to counsel their student for lack of motivation or work ethic.

As for counseling and mentoring, follow the simple rule: Talk to someone like you would like to be talked to. If trainers take this approach and understand that students are trying to improve (and that some will improve at their own level or tempo), then the trainers will not get too frustrated. Also, remember that, in the end, it is up to the students to improve. The amount of effort and time they put into practice is what will get them to the next level.

## TEAM AND ORGANIZATIONAL TRAINING TIME MANAGEMENT

It would be unrealistic for me to specify that individuals get this much training time, teams get this much training time, and the organization gets this much training time.

The first building block of training is the *individual.* He must master the safe use of his weapons to be able to train with a team element. He must be in shape and focused and have all his equipment serviceable and ready. Who trains the individual? Several people have influence here. First, his initial academy or training section does. This academy or training section may work for the operational elements and may be checking to see if the product (individual) is meeting the needs of the team or organization. When I went through Special Operations training, the staff was able to get us to a 50 to 75 percent level of a seasoned soldier. The receiving group or team was responsible for "polishing" and getting us to the level we needed to be able to support and effectively bring the team to operational levels.

Years later, when I worked as an instructor in this section, my goal was to bring the students to an even higher level so when the teams received them, they were closer to operational than I had been. It was my measure of success in the job. I felt if I bettered my classes, which in turn bettered my students, they would have an easier time fitting in on the team and their learning curve would not be that steep. This relates to law enforcement in that the academy graduate should not have a great learning curve when it comes to working on the street. The field training officer should not have to retrain a rookie to perform the job. Minor "tweaking" is all the rookie should need.

Training cells have a responsibility to check with the operational units to verify requirements standards. It is also the responsibility of the operational elements to direct the academy to modify or add classes to better address problems on the street. In the end, the academy or training cell should be a support element responsive to the needs of the operational elements.

The *team* must master tasks and subtasks. A task might include moving a team to a breach point and solving problems on the way. A subtask might be handcuffing a suspected threat. This includes actions at breach points, hallways, CQB, etc.

The *organization* is responsible for organizing, moving, and employing several teams on a target at one time, then collecting all the information the teams and processing the target in a logical sequence. This includes all the associated problems such as officers down, injuries, etc. This also includes running the TOC and disseminating critical information in a timely manner.

## EQUIPMENT MANAGEMENT

Before I get too deep into equipment management, I want to discuss a few items. The first is the human factor. Know what heat and equipment do to the human body and how these elements affect a student. The easiest way to gain this knowledge is to wear the same gear as the student. If wears the instructor out, it is having the same impact on the student. If the instructors are thirsty, so is the student. Instructors will quickly realize when to take breaks and make sure the students do the same.

*Dry*, *slick*, *gear*, and *wet* are all terms that I learned years ago. As noted earlier, dry refers to dry runs which involve using weapons that are unloaded where I am teaching the physical and mental muscle memory necessary to execute the drill. Slick means without gear or with weapon only. This way a student can focus on the task and not the environment factors. Sometimes, I can only get so much out of a student's human body and I want to save the students for live runs. This could be a place where the temperatures are too extreme to run all dry practices in full gear. Students may not be acclimated. I have been deployed from a wintertime U.S. environment to areas that were tropical and hot in a matter of hours. A body simply needs time to adjust. Gear or equipment is just what it means, everything down to gloves. Once the students are proficient and safe, they can go *wet* or *hot* with live ammunition.

When conducting training with several modules, you can use a single instructor for a single block of information, or one comprised of a lead instructor and assistant instructors for a large group of students. If I have multiple blocks of information that don't require a training sequence, I can use several instructors in a round-robin type of training environment.

## YEARLY AND REPETITIVE TRAINING

I discuss this topic earlier in chapter 7, but it is so important it bears repeating. Often referred to as Maintenance Training, we must continually practice and master our skills as they are perishable. That is, our proficiency at executing them smoothly will degrade with time and loss of practice. Teams need to review entire programs every few years due to turnover of personnel or infrequency of the mission.

What ratios should teams be trained to? This is hard to say. Operationally, if you perform a mission like raiding targets and you are going through the planning and execution on each of them on a weekly basis, you probably do not need a maintenance training session on them. In the law enforcement arena, if an officer is processing several DUIs a week and is keeping his skills up, he probably does not need an organizational class on it. I can defer the individual training to the team leaders to polish up anything they see as being a weakness or deficiency.

If students practice a particular mission such as Hostage Rescue only once a year, it would probably be better to practice this mission as a whole unit/element at least twice a year. As an organization can schedule monthly training in a manner such as this:

| Month | |
|---|---|
| 1 | Hostage Rescue |
| 2 | High-risk Warrants |
| 3 | Search Warrants |
| 4 | Barricaded Persons |
| 5 | Vehicles/Buses |
| 6 | Evaluation and Remedial |

As discussed, an organization could assign a specialty team to plan, rehearse, and run the training if a large training staff is not available. A good guideline is train for two days each month. Give the teams a half a day for individual training, another half day for team training, then half a day for organizational training.

The key to keeping it simple is to create tactics and techniques that can be used on all missions. For example, the technique of approaching a building either on foot or via a vehicle can also be used for Hostage Rescue, High-Risk Warrant, Barricaded Persons, etc. This way the team's plans are simple and the lowest-level member of the team (every team has at least one Gumby) does not have to remember a great deal from month to month.

This thought process applies to Close Quarter Battle, Breach Points, Collapsing a Target, etc. Having a different system for all missions is good for nothing except setting the team up for failure.

Active Shooter training is another area in which many departments perform the training once and do not do it again for years. This is a highly perishable skill and it should be a yearly training requirement. A subcomponent of this training is medical training, and this should also be worked on at least once a year.

## KEY POINTS

- Keep continuity of instructors and information to include lesson plans and video of the classes.
- Use specialty teams to help teach tasks to the larger organization when possible.
- Understand that diagnostics take up valuable time and students will need to be retrained in any case.
- If an individual does not cut it, do not put them in a training capacity where new students will see them as a possible example to follow.

# 19

# MANAGING HIGH-RISK TRAINING: THE TACTICAL RIFLE AND PISTOL INSTRUCTOR MODELS

*If it is worth looking at, it is worth looking at with a gun.*

- **GENERAL OVERVIEW: TACTICAL RIFLE INSTRUCTOR COURSE**
- **CLASS SEQUENCE**
- **SAFETY PROCEDURES**
- **INSTRUCTING CLASSROOM MODULES**
- **PRACTICAL EXERCISES AND TRAINING**
- **TEAM MANAGEMENT**
- **REHEARSALS**
- **USE OF VIDEO AND AAR**
- **FINAL SCENARIOS**
- **A LITTLE HISTORY AND HOW THE COURSE EVOLVED**

## GENERAL OVERVIEW: TACTICAL RIFLE INSTRUCTOR COURSE

Now that I have explained the general concept regarding high-risk tactical training, I want to explore the most common courses instructors and students are likely to encounter during their careers.

---

### DAY 1

*AM*

| | |
|---|---|
| • Rifle | |
| • Standards | |
| • Safety | Yellow Cord/2 Steps Back |
| • Fundamentals | |
| • Zero | 7 Yards |
| • Dry Fire | 100 Yards |
| • Prone Confirm | 100 Yards |
| • 3-3 Round Groups | 100 Yards |
| • Kneeling 5 or 3/1 Drill | 75 Yards |
| • Kneeling 3-3 Round Groups | 75 Yards |
| • Kneeling 3-3 Round groups | 50 Yards |
| • Standing | 25 Yards |
| • Discuss Shooting Plan | Index Card |

*PM*

| | |
|---|---|
| • Speed Shooting | |
| • Ready position (1 Mag—Group/Ind Time) | 7 Yards |
| • Transition (Dry/Group/Ind Time) | 7 Yards |
| • 2 Shots—One Target (1 Mag) | 7 Yards |
| • 2 Shots—Two Targets (1 Mag) | 7 Yards |
| • 5/1 Drill (Group/Ind) | 7 Yards |

DAY 2

*AM*

| | |
|---|---|
| • 100 Zero | CSAT |
| • Barricades Individual (4 Positions) | 75 Yards Steel |
| • SPORTS | 75 Yards IPSC |
| • Reloads | 75 Yards IPSC |
| • Standards | CSAT/Shirts |

*PM*

| | |
|---|---|
| • Corner Clears | |
| • Maintenance Class | |
| • 100-7 Yard Drills | Bulls/IPSC/Shirts |

The above course is expanded over four days to allow more time to discuss methodology. I have the class shoot the standards at the beginning of day 2. Between shooting standards, modules are added to give students a physical and mental break from qualifying. Before each standard, the students shoot remedial courses of fire. This allows time to work on the drills that most of the class failed and helps more students pass their shooting qualification.

## INSTRUCTOR TEACHING ASSIGNMENTS

Day 1

| Class | Instructors | Remarks |
|---|---|---|
| Introduction | PAUL | |
| Zero and Confirm | | |
| Prone Position | | |
| Kneeling Position | | |

*LUNCH*

Standing Position
1 Shot
Transition
2-Shot Drill
6-Shot Drill
2 Shots/2 Targets

*IF TIME PERMITS*

Standards TEAM #1-PD RANGE
Standards TEAM #2-INSTRUCTOR RANGE

DAY 2

Team #1
Barricades Tac Positions/Reload/Malf
Shirt Drill Record Times Only

Team #2
200/300 Shoot 2–3×
Scrambler

*EARLY CHOW*

100-7 Drills Instructor Range
Surgical Movement Drills
Standards

100-7 Drills PD Range
Surgical/Movement Drills
Standards
Weapon Maint. Class Paul Police Range
Certificates Paul Police Range

## CLASS SEQUENCE

The Tactical Rifle Instructor course is a six-day course. Day 1—4 is the instructor training portion and day 5—6 culminates in a two-day Tactical Rifle Operator class where student instructors, under my supervision, put on a two-day class. Instructor students are required to give one, or several, blocks of instruction along with managing safety and running the line.

My intent for the instructors at the completion of this course is to be capable of safely running a two-day operator course. In general, I put instructors through the two-day course in day 1—4, but at a slower pace with more emphasis on how to, and why, I do what I do. Instructor students see the class one time in this phase, are assigned blocks of instruction, and must study for and rehearse any demonstrations for any assigned instruction they are teaching during the two-day class. In essence, they see the information again. Finally, they see the information a third time in the two-day operator course and flow.

I begin the class with a talk on why zeroing procedures are performed at the seven-yard line. I use a marker to draw sight pictures on the targets of how students should be seeing their sights or optical dots. I show students load/unload procedures so students begin with the proper base. Students fire a couple of strings at seven yards to get their baseline zero before the class moves back to one hundred yards for our battle sight zeros. Students unload and put chamber blocking devices after every time we unload to build muscle memory. Also, when making sight corrections, I want a visual aid to ensure the guns are unloaded.

Once students have a baseline zero, I send them back to the one-hundred-yard line to top off their magazine and standby. Once I get everyone off the seven-yard line, I get the class together and go over the prone position I want them to shoot.

I demonstrate a drill, and then have students practice it. This way, students avoid information overload. I first begin with dry practice on the line where I can fix the major position problems. Once students start shooting, I start working on their shooting mechanics and trigger pulls.

The class keeps shooting until all students have moved from the bull zero target to the CSAT silhouette target. This way I can monitor the progress of the class. Normally, I have students shoot three-three round groups at the center bull and then make a sight adjustment. Once students are zeroed, they transition to the CSAT silhouette target that the class uses to shoot standards. Normally, students begin firing in the prone position. As students move to the CSAT target, I have them start shooting from the standing position, moving to the prone. Students will then shoot the drill for time.

Some students zero early and some take a few relays. In any case, all students are constructively engaged and building their skill base. Instructor courses only differ in that I take the time to explain more of the methodology and rationale behind the drills. Day 1 normally follows the same for instructors and students except for a bit more lecturing in the instructor class.

The instructor class then shoots the standards for score. As instructors qualify, they stop shooting and begin to run the line. Instructors learn to work timers, operate the speaker system, and record scores. Students learn to issue commands as a range master. Each student who qualifies progresses the same way. After the class shoots the standards, I change the students' focus and introduce a module to them.

An example of one such module (or block of instruction) would be barricades. This module is identical to the module an instructor would teach to new students. Instructors learn kneeling right and left and standing right and left while shooting at about eighty yards on steel chest plates. Once students do this dry, then live, students reload magazines and discuss tactical reloads. This is a natural progression because students are behind cover and learn to take advantage of it.

I use this same procedure for standards and modules until day 3 of the course. On day 3, I begin back briefs of assigned classes for the two-day student class. I want instructors to begin rehearsing so the two-day student class is a seamless operation. If the class requires demonstrations, I want to see instructor students perform the shooting demo. I want their demos slow, deliberate, and mechanically perfect. Students will remember the last thing they see. If an instructor performs a drill substandard, students will emulate it.

## SAFETY PROCEDURES

I use the same safety sequence for all my shooting classes, and this begins when students first start on the firing line. I provide students with a written handout for review, but simply practice the same verbiage with each firing order to ingrain simple and easy-to-understand verbiage. Simple terms that students must comply with are *load* and *unload*. Students are given a load-and-unload procedure and are checked to ensure consistent use. While students are shooting, I try to keep the verbiage simple and correct only one problem at a time. I pick the worst common error students are making and between strings remind them to perform the sequence correctly.

I start my verbal commands with, “Eyes and ears.” I want to ensure that students have on ear protection and eye protection and that they are not doing this with a loaded gun in their hand. Shooters have died adjusting earmuffs with a loaded pistol. No kidding.

My training sequence for all demonstrations that may involve high-risk is to dry-fire or dry practice first. This way, I save time and ammo by correcting students before they have to fight the weapon recoil. This is easier for their minds to absorb since their attention is not diverted to an explosion at the end of their gun. I can watch students conduct the drills dry and fix any major problems or safety issues with minimum rounds being fired. This is where I use the 5/1 drill. Five dry fire to one live-fire. They fire a total of five live rounds but get thirty practices in at the same time. I developed this drill when I left Special Operations and could not afford to shoot the same amount of ammunition I was used to. I would make a box of fifty last for two hours and get the same benefit of firing several hundred rounds (for more information on 5/1 drill, see chapter 15).

## INSTRUCTING CLASSROOM MODULES

When it comes to tactical training and firearms in particular, I cannot emphasize the need for field training enough. I limit classroom time in this course due to the need to learn and practice the skills on the range. I have seen other instructors spend an inordinate amount of time in the classroom describing something that can be taught in two minutes on the range. The extra time used in the classroom can be put into building muscle memory on the range with dry- and live-fire practice. Some instructors simply prefer the classroom to the range because of their comfort level or their inability to shoot or demonstrate drills. I want classes to learn to shoot on the range. I use the classroom when weather turns severely bad and learning is not productive on the range.

## PRACTICAL EXERCISES AND TRAINING

I prefer to demonstrate all practical exercises with all my equipment and gear. I wear all my tactical gear when teaching because it changes the way I shoot, address the gun, and is difficult to master. If I can teach a student to shoot comfortably in all his equipment, I can then teach a student to shoot in a patrol officer's uniform, a concealed carry mode, no gear, etc. My idea is to have one stance, grip, etc., that will work in all the above systems. Why waste time trying to teach students a system that only works in a single situation? I prefer to teach one technique that can be used in all situations and assignments. It will save students time in going back and forth in multiple systems and they can focus their time and energy on one. Remember training time and ammo are valuable commodities. Why teach an officer three or four different systems over the course of their career?

Another important point of wearing the gear is so that students witness that I can perform the drill myself. They also observe how I rig and structure my equipment for success. Also, by wearing my equipment, I can monitor the fatigue level of students and ensure they are not being placed in unsafe heat or cold without proper breaks or hydration. Finally, in the end, students will not look at you and wonder why you are not in the same uniform.

## TEAM MANAGEMENT

As I have stated before, I prefer to teach leadership in every course and I use strong students to help run team concepts. I like to break down large classes of twenty to twenty-four into two small groups of ten to twelve for ease of control and to keep them constantly moving and occupied. By using a team concept, teams get to know their instructors and vice versa. It works well.

I assign instructor students into two teams and each team manages ten to twelve students. In the end, the instructor to student ratio is 1:3 and I have instructors work with the same two to three students to build rapport and to keep the instruction consistent. Students will get tired of multiple instructors telling them ten different things. It is easier for a single instructor to work with them and fix one problem at a time.

## REHEARSALS

Military and law enforcement personnel are used to rehearsals, the act of practicing a drill, a teaching point, or a demonstration. The purpose of a rehearsal is to practice before performing to ensure the student instructor (SI) meets the satisfactory standard. This is where I have had the most problems with civilian students.

Generally, once SI students pass the shooting qualification and written test, the SI students feel that they have completed the course. They take lightly the need to deliver a superior class to the new students during the operator training course.

What SI students do not know is that rehearsals began when I started teaching. I use the same techniques, verbiage, and range commands as I want used when they teach the course. Most cannot remember and have to be retaught. I use a natural progression in my instructor courses where when you pass the shooting qualification, you begin running the line. SI students get practice running a clipboard, scoring, and the use of the shooting timer. SI students also get to practice issuing verbal commands and explaining drills. SI students practice giving range commands to a group. I use a portable public address system and SI students must learn how to operate it. In regard to lecture I teach that "less is more efficient."

Many SI students do not like to talk and get tongue-tied. I use simple words like *load* and require all students to go through a preprogrammed sequence to get their weapons ready to shoot. Another problem SI students encounter is that they want to push the students they are teaching, too fast. As SI students, they are required to pass time standards in order to graduate.

When SI students teach new students, I shift the emphasis for speed to perfect mechanics. I want students to shoot only as fast as they can accurately. I use the scoring box on my CSAT target to help students shoot fast and accurate. If a student starts pushing rounds out of the box, they know to slow down and they have hit their maximum speed level for the time being. Students must work on their mechanics to improve their speed. I refrain from mentioning time standards when SI students, or I, first teach a new student a drill because their minds focus on beating the clock versus performing the drill correctly.

## USE OF VIDEO AND AAR

I have started to use video in a Tactical Rifle Instructor class as it allows students to see their head-eye relationship which needs to be consistent for accuracy. I also use video extensively in the pistol portion at the end of day 1. I video both SI and new students performing five one-shot drills from the high ready, five from the holster and three five-shot strings in the chest and one in the head failure drills.

I have students come to the line with an instructor behind them providing range commands and shooter's times. I usually video from the side or at a slight angle where I can see the entire shooter, head to toe, then zoom in on their upper body and gun. I video the first couple of shots looking at the whole shooter to see stance, weight, etc., then zoom in to catch the upper body and gun. I am looking to see if their finger is popping off the trigger, if they are aligning their sights for a second shot, and if they perform a proper cover or scanning procedure. During the debrief, I use a laser pointer to pick out points and rewind and slow down from time to time.

I use the same procedure with the students shooting from the holster. I want to see if the holster is positioned too low or too high on the belt or if the shooter is dipping a shoulder on the draw. I want to see the shooter's draw stroke, off-hand position, presentation, and shot sequence. I also want to see them reholster and if it is done safely.

Finally, I video students firing a 5/1 drill, five to the body and one to the head as fast as students can see their sights during firing. I use this drill to check student's overall balance and if their arms are locked during the firing sequence. I do this by looking at them overall and how a student's head moves against the background or skyline during multiple shots. You can easily see if students are being pushed out of position. If students get into a more aggressive position on the 5/1 drill from the start, students should be doing this on all of the drills they shoot. One never knows how many rounds they are going to need to put the opponent down, so students should be aggressive whether it be a one-shot, two-shot, or 5/1 drill.

I ask students to pick out two to three items to work on during this day. Again, the reason I critique in the morning is because students are fresh, rested, and open-minded. They are ready to better themselves and have a full day to do it.

## FINAL SCENARIOS

The final test for my instructor students is to deliver a block of instructions to new students, run the line, and provide feedback and correction. They do this under my supervision on day 5—6. Generally, by this point they are all tired from four days on a flat range. My job as lead instructor is too keep the class going to standard and to correct any misinformation or information that has been left out. I owe it to the instructors and to the new students. I add points after an instructor gives his material by saying, "I would like to add a point." Instructors understand this is not a critique of their material, but a quality control of sorts.

When I let students out on each break, SI students conduct a "group hug." I give them a quick overview/critique of instructor performance. I also point out anything they need to do as a group to polish up and attempt to accomplish this in a positive manner and not close SI student minds on the last day. I must have extreme patience as these SI students are still new students.

## A LITTLE HISTORY AND HOW THE COURSE EVOLVED

I began teaching instructor courses as six-day modules as I never believed a SI student would learn as much in a short course without teaching a new student and presenting blocks of instruction.

I added several drills that do not give too many tactics away, but rather than make both the SI and new students safer in their use of the weapon. I now include a surgical and movement drill. I put up five surgical targets on a flat range with a space in between. I use both SI students and new students as "door stops" facing on line in the spaces on the seven-yard line. I require students conducting the drill to move from a start point around each student safely with the weapon to a firing position between students and to engage each surgical target with one round. The targets are five heads with full exposures to one-quarter head exposure next to a hostage. Students shooting fire one round per target and do this five times. Students must know the line-of-sight–line-of-bore relationship when shooting with the rifle and must perform a good lock out with a pistol to ensure they do not hit the hostage targets.

As for target progression, I show students several targets. I use bull targets for zeroing with the rifle, then move to a CSAT silhouette target. I have students shoot the standards on IPSC targets inside a shirt. This teaches the student to pick a spot on the shirt and drive their sight to that same spot. Next, I use the surgical targets described in the previous paragraph. I want students to know the capabilities of their weapon and their own skill level. Additionally, I use the bull targets for distance and grouping work at twenty-five yards. With three bulls, I can work ready position, holster, and kneeling on one target and identify each group.

## KEY POINTS

- Have a logical sequence and structure for each course.
- Make the course positive and try and help students absorb required material.
- Shoot the standards with the students at some point to keep yourself honest and show students how it can be done.

# 20

# MANAGING HIGH-RISK TRAINING: SHOOT HOUSE INSTRUCTOR

*Whole person, hands, waist line, immediate area and demeanor.*
—Discrimination Process

- **GENERAL OVERVIEW: SHOOT HOUSE INSTRUCTOR**
- **SAFETY**
- **INSTRUCTING CLASSROOM MODULES**
- **PRACTICAL EXERCISES**
- **TEAM MANAGEMENT**
- **FINAL SCENARIOS**
- **A LITTLE HISTORY AND HOW THE COURSE EVOLVED**

## GENERAL OVERVIEW: SHOOT HOUSE INSTRUCTOR

When I developed this course, I ran into the dilemma of whether to teach CQB or not. Thinking about my end goal, I quickly determined that I would need to teach a system of CQB for safety and also for consistency of instruction and management. Taking it a step further, I would have to give students a crash course in Pistol and Rifle Marksmanship and Weapon Handling. It would become a busier course than I had planned.

---

## SHOOT HOUSE INSTRUCTOR (SHI) COURSE CHECKLIST

### SHI HANDOUTS/PREP

- Packets
- Safety Letter
- Flat Range AM (24 Targets, Shirts, Bulls, Ammo, Etc.)
- 4× Knock-Down Target Stands—Cover Drills
- 2× Knock-down Targets—Corner Clears
- Charge Video Batts
- Charge Drill Batts

| *DAY 1* | *SHI TEACHING POINTS* |
|---|---|
| • Introduction | |
| • Group Introduction | |
| • Roster | |
| • Team Breakdown | |
| • Week's Schedule | |
| • SAFETY | POWERPOINT/VIDEO |
| • SHOOT HOUSE OVERVIEW | POWERPOINT/VIDEO |
| • CQB OVERVIEW | POWERPOINT/VIDEO |
| • SURGICAL SHOOTING | POWERPOINT/VIDEO |
| • RANGE FIRE RIFLE/PISTOL | |
| • IF TIME PERMITS | |
| • Tactical Flare | |
| • Corner Clear | |

- Cover and Hands
- QUIZ #1 REVIEW

*HANDOUTS/PREP DAY 2*

- Targets
- Video Gear

*DAY 2* *1–2 CQB DRILLS*

- Safety
- QUIZ
- 1–2 PERSON CQB POWERPOINT/VIDEO
- CQB TRAINING METHODOLOGY POWERPOINT/VIDEO
- RANGE: TEAM SETUP 1/2 AND 3/4
- PRACTICAL EXERCISE 1 AND 2 CQB
- CENTER/RIGHT/LEFT/IMM THREAT

*HANDOUTS/PREP DAY 3*

- Individual CQB Checklist
- Knock-Down Targets
- Hands/Guns
- Timers
- Targets

*DAY 3* *TEAM CQB LIVE FIRE*

- QUIZ
- Review Video
- Safety
- TEAM CQB POWERPOINT/VIDEO
- FLASH BANGS POWERPOINT/VIDEO
- PRACTICAL EXERCISES
- FATIGUE INDICATORS
- MEDICAL (IF TIME PERMITS)

*HANDOUTS/PREP DAY 4*

- Breaching Tools
- Dummy Flash Bangs

*DAY 4* *MULTIROOM CQB*

- Review Video
- Safety
- QUIZ
- MULTIROOM CQB POWERPOINT/VIDEO
- ASSAULT BREACHING POWERPOINT/VIDEO
- LOW LIGHT POWERPOINT/VIDEO
- MULTIROOM PE
- 1 TEAM SETUP/3 HIT

*DAY 5* *TEST*

- SHI INPECTION
- 1–2 PERSON CQB SETUP/EVAL
- 3–4 PERSON CQB SETUP/EVAL

---

The Shoot House Instructor course is a five-day, forty-hour course designed to teach how to safely manage a live-fire shoot house and how to teach Close Quarter Battle (CQB) to new students. The above informal checklist is used to ensure I am covering all the required material. It also allows me to prepare equipment for the next block of instruction the day before so there is no stand-around time.

*Day 1*

Day 1 begins with the routine administrative tasks and paperwork. I then divide the class into four teams of five persons each and assign a TL. My reason for this goes beyond administrative requirements, but moves into the tactical and learning arena. I want the teams to become used to working around each other when performing live-fire CQB. Switching students between teams in a large class fails to build cohesion. Students may not become as proficient as they would while staying with the same team.

Due to the amount of information in the course, I provide my routine one-hour safety briefing and an overview of CQB, two-person and with teams. I use PowerPoint programs imbedded with video to show how it should be done and the speed students can attain with practice.

After lunch, the class meets at the range. I start with pistol and cover basic drills that should be used in the house. I show the proper load and unload procedures and the proper shooting stance with tactical gear. Students shoot at least five of my ten standard drills to work on marksmanship and safety fundamentals. I also show the *sul* or "depressed ready" stance with the pistol. *Sul* comes from the Portuguese language for "south." The students need to practice depressing the muzzle when fellow students or instructors were in front of them.

Once they finish with the pistol, I move on to the rifle and cover roughly the same drills. I start with a load-unload sequence and move on to the stance and one-shot drill. Students worked these drills from the low-ready position where the muzzle is depressed below the student's beltline. This enables students to have a clear view of the room and to discriminate. They are actually faster in this position than trying to run in with their rifle stuck up in their face, which is like trying to look for multiple targets for a gun through a paper towel tube.

Students cannot see the big picture in that position, let alone obstacles in their path that they might stumble or trip over.

The class finishes up the day firing a few rounds of frangible ammunition to ensure the weapon function and the accuracy is acceptable. Once this is accomplished, the class builds targets for the week and performs a quick walk through the shoot house.

### *Day 2*

Day 2 begins with a short quiz in the classroom. Students review the two-person CQB class from the day prior. I provide a class on CQB instruction and methodology that shows the priorities and how to safely stair-step training. At the conclusion of this class, teams are separated into a CQB team and an instructor team. This effectively divides the house into two lanes of three rooms each. I assign one side Pistol CQB and the other Rifle CQB. Instructor teams set up their respective sides of the house and prepare for dry runs with their sister student teams.

After the initial setup of the house, I demonstrate students how to inspect rooms for safety using a checklist which I developed and provide to the class. CQB students are told to report to the unload point and clear their weapons. Team leaders managing CQB are told to check their assigned rooms and report to me for inspection. Student instructors then run their students dry through the house.

These dry runs give students a chance to practice their CQB techniques in the same rooms on targets and train them what to expect. The instructors practice inspecting students outside the door, entering, and moving to their point of domination. Instructors learn how to move in and around students and obtain a second look at the room. This second look allows instructors the ability to reevaluate the safety aspects of their rooms and adjust anything that might create a potential problem such as target placement or furniture.

Once student instructors feel their students have mastered the two-person entries, the house is declared "hot" and students are told to move to the load/unload point and load. Additional student instructors are normally stationed at the load/unload point should the requirement of frangible ammunition be needed. Many shooting houses require the use of disintegrating ammunition in the facilities.

Student instructors, not needed on the ground, are issued video cameras. They are directed to the catwalk above the house and video record student live runs for review at the beginning of the next day. Instructors on the ground then run their students through live-fire drills.

### *Day 3*

Day 3 begins with a short quiz, then video review of day 2 and time for any questions about two-person CQB. The class takes a break before reviewing the team level CQB PowerPoint. After this PowerPoint, I teach a short block on how to deploy flash bangs in conjunction with a team performing CQB. Once this is complete, the teams are again divided up and assigned to different sides of the shoot house to run their sister teams in dry team-level CQB. The sequence for day 3 is the same as day 2.

### *Day 4*

Day 4 begins with a quiz and video review of day 3. I answer any student questions and move on to multiteam/multi-breach-point exercises and how to set them up.

A PowerPoint presentation is shown on how to conduct this type of training and conduct "tabletop" exercises on how to setup houses which have bullet traps and bullet trap walls. The final exercise is conducted in the shoot house. One team is assigned to set up the house and manage the three teams entering the house, looking at linkup points, limits of advance, etc.

As with the other classes, dry runs and inspections are run until students know their jobs and areas of responsibilities. Students run the entire scenario and the student instructors offer their critiques. Another team is assigned to set up the house for the rest of the class until every team has completed the scenario.

### *Day 5*

Day 5 begins in the shoot house with the inspection of a room by each instructor. I set the room up for first time CQB students to include deficiencies (unsafe target placement and situations). I require instructor students to come in and inspect the room I have set up and within two minutes give me three deficiencies. Generally, there are about ten, but I want to ensure instructor students have a system for looking at a room.

When this is completed, I assign student instructors to set up rooms and begin conducting dry CQB training. They later conduct live-fire CQB modules with assigned students. I watch instructors managing training and students performing it to ensure they can do both, manage and perform CQB.

## SAFETY

The Shoot House Instructor course sounds like a high-risk course. It may be to some, but in general, it is one of my safest courses to teach. I structure the classroom and practical exercise flow to quickly build on the previous block of instruction. This building process ensures students do not move to the next level without passing the previous level.

With the theory of showing students the "whole elephant," I show the entire CQB system to give students the understanding of where they are going and how they are going to get there. Once students see the entire pie, we begin to eat it one slice at a time.

## INSTRUCTING CLASSROOM MODULES

I use PowerPoint presentations heavily embedded with video that I have collected and choreographed over the years. Other training aids I have used are "blue" or "red guns" fitted with lasers and pressure switches. Using the existing doors of the classroom, I have students enter and move to their points of domination as a demonstration. Using the laser, I point out where their sector begins and ends. This way all students see where their muzzles start and finish, ensuring they do not point weapons at fellow officers.

## PRACTICAL EXERCISES

When the class moves to the shoot house, I begin with students setting up the rooms per the checklist. I then perform a walk-through and a room inspection with the entire class. I ask for a "dream team" or demo team to use when walking through the room clearing process.

I start with two students from the demonstration team stacked up on a door. I talk about what I look at as an instructor when I scan students for safety before they enter a room. I start at the head and work down. Eyes, ears, head posture, weapon posture, trigger finger index, mechanical safety, etc. I scan both students in the same manner to ensure their preentry posture is correct. I have students move into the room at one-fourth the actual speed to talk them and the class through what they should be seeing and doing.

Verbally cuing them with the sequence helps them to understand what to do and why they are doing it. If there are no questions, I turn the students over to their student instructors for dry runs. I monitor all student instructors and their dry runs from in the shoot house, moving from room to room and instructor to instructor. Once student instructors are satisfied with their students' performance, they are clear to go with live ammo.

## TEAM MANAGEMENT

I continually hold TLs accountable and require them to run their teams and set up. I usually stack the deck with hard-charging TLs who set the standard from the start and set a high mark for all others to strive for. If I put a weak TL in a position at the start, I generally will have poor standards set and will spend valuable time correcting them. It causes more work and sometimes causes confusion for the teams.

Once the standard has been set, I rotate TL assignments daily and sometimes two times a day. I encourage the TLs to work as a team and help the less experienced out.

## FINAL SCENARIOS

The final day consists of a comprehensive review in the shoot house. I begin with a simple room inspection and ensure that students know how to conduct safety checks and how to spot dangerous situations and correct them. SI students review both CQB and shoot house setup and management. I begin by choosing instructors to setup and train basic scenarios and move to more advanced scenarios. All student instructors must perform their assigned tasks as instructors and as students performing CQB.

Time permitting, I expose students to advanced scenarios that they may have not seen during the week. I want to keep adding to their knowledge base. In the end, I ask them if there are any scenarios that I did not cover or if they have questions.

## A LITTLE HISTORY AND HOW THE COURSE EVOLVED

As the course evolved, I saw key issues come to light and adjusted for them. The first was the fatigue factor and safety. Where an experienced CQB instructor might teach several days in a row, new SI students are not up to that level and this must be taken into consideration. Some SI students are "range masters" who inherit shoothouses and need to safely understand how to administer them. They may not have the tactical background or CQB experience a SWAT officer possesses. With that, they are trying to absorb a great deal more information to apply it at one time. They will fatigue faster. This is why I break the groups into shoot house managers and students, and change duties often.

CQB is more of a mental game than a physical one. Yes, wearing all the equipment will tire students if they are not in shape. The mental fatigue comes into play since every room entry requires precision and total focus at all times to ensure each room is safely processed. As the mind tires and students become accustomed to the targets, I begin to change the gun to a different hand and put badges on targets. In active shooter response scenarios, officers may come in every type of uniform imaginable, including civilian clothes, both well dressed and undercover. During the carry mode, most police officers keep their badge next to their gun in case the gun is seen, then the badge is also seen. The problem is that when the gun comes out, the badge is still attached to the belt. I want officers to be able to see the badge before they drop the hammer and take a human life.

Next, I have targets with intertwined arms that look like an octopus. The student must consciously slow down and see the gun and follow the hand to the arm to the body to see whom it belongs to. Failure to do this will result in the wrong person being shot during a close-in physical altercation.

"Getting strong for doing wrong" is a phrase used for payment when a hostage is shot or rounds miss a target. It happens, but it needs to be reinforced that it was either poor judgment or sloppy marksmanship. Generally, for every round not where it was supposed to go, twenty-five push-ups are required. If it is done with a two-person element, both people pay. If it is done with a team, the entire team pays. It sends a message home that the student needs to tighten up his work. Failure to do this will result in a careless or carefree attitude that will carry over to the street. This is unacceptable.

Finally, all rooms are checked and double-checked prior to CQB. This means if an instructor sets it up, he checks it, then has another team member check it or another TL. I have seen too many shoot house errors caused by poor instructor set up resulting in students getting shot by other students. It only takes a moment to check, so take the time.

## KEY POINTS

- When conducting live-fire courses with a new group, always review marksmanship fundamentals and safety before going into the shoot house.
- Use internal checklists when needed to insure that you are ready for the next day. This limits downtime.
- Always review information taught during the week on the final day.

# 21

# MANAGING LEADERSHIP TRAINING: THE TACTICAL TEAM LEADER

*Leadership in all things.*

- **GENERAL OVERVIEW: TRAINING THE TACTICAL TEAM LEADER**
- **SAFETY**
- **TACTICAL LEADERSHIP**
- **COMBAT MIND-SET OF A TACTICAL TEAM LEADER**
- **LEADERSHIP COUNSELING**
- **SELECTION AND TRAINING**
- **PRACTICAL EXERCISES**
- **ASSAULT SEQUENCE**
- **ASSAULT PLANNING AND OP ORDER**
- **REHEARSALS**
- **COMBAT LEADERSHIPS MODULES**
- **TARGET PLANNING BRIEFINGS**
- **A LITTLE HISTORY AND HOW THE COURSE EVOLVED**

## GENERAL OVERVIEW: TRAINING THE TACTICAL TEAM LEADER

The Tactical Team Leader course is a three-day, twenty-seven hour course designed to teach TLs and future TLs how to effectively train, manage, and lead tactical teams.

While developing this course, I ran into the problem of how long to make this course. It is difficult for leaders to get time off from work because the higher you get up the food chain, the more responsible you are. Hostage Rescue is used as the main course focus to train on since it is the most difficult mission to perform. Adding the missions of Search Warrant, Barricaded Person, and High-Risk Warrants would be too much information to cover. Instead, I try to keep principles simple and use one medium for training.

---

| DAY 1 | 0800 | 0815 | Introduction | PP | Classroom |
|---|---|---|---|---|---|
| | 0815 | 0830 | Safety | PP | |
| | 0830 | 0915 | Team Breakdown/ Organization | PP | |
| | 0915 | 1015 | Tactical Leadership | PP | |
| | 1015 | 1115 | Combat Mind-set | PP | |
| | 1115 | 1215 | Leadership Counseling | PP | |
| | 1215 | 1300 | Lunch | | |
| | 1300 | 1400 | Selection and Training | PP | Classroom |
| | 1400 | UTC | PE #1 Selection/Training Module | | Field/Range |
| | | | 1-Mile March - Tactical Gear | TL 1 | Range |
| | | | -Long Gun Shoot - 100 Meters | TL 2 | |

| | | | | | |
|---|---|---|---|---|---|
| | | | 2-Mile Run—Pistol Belt | TL 3 | |
| | | | 25-Yard Pistol Shoot | TL 4 | |
| | | | -PT Test (PU/SU/1-Mile Run) | TL 5 | |
| | | | (Or Team O Course) | TL 1 | |
| | | | AAR | | |
| | | | Training after Action Briefing | | |
| | | | AAR Format | | |
| | | | Issue Blue Cards - Self-Assessment | | |
| DAY 2 | 0800 | 0845 | Assault Sequence | PP | Classroom |
| | 0845 | 1045 | Assault Planning and OP Order | PP | |
| | 1045 | 1115 | Tactical Thought Process | PP | |
| | 1115 | 1145 | Rehearsals | PP | |
| | | | Issue Tactical Planning Packets | PP | |
| | 1200 | 1300 | Lunch | | |
| | 1300 | 1500 | Combat Leadership Modules: | PP | Classroom |
| | | | Safety | PP | |
| | | | Premission | PP | |
| | | | Emergency Assaults | PP | |
| | | | Movement | PP | |
| | | | Breaching | PP | |
| | | | Multi–Breach Points | PP | |
| | | | Anti-Fratricide | PP | |
| | | | CQB | PP | |
| | | | Medical | PP | |
| | | | Hostage Handling | PP | |
| | | | IEDs | PP | |
| | | | Consolidation and Exfiltration | PP | |
| | | | Post Assault | PP | |

| | | | | | |
|---|---|---|---|---|---|
| | 1500 | 1700 | Tactical Training Modules | Hands On | Target |
| | 1700 | 1800 | Dinner | | |
| | 1800 | UTC | Team Planning Session and Briefings | | Classroom |
| DAY 3 | 0700 | | SCENARIO BRIEFING | | Unknown |
| | 0800 | | #1 DELIBERATE ASSAULTS<br><br>W/MEDICAL<br>2 Hostages - 1 Injured<br>1 Iteration per Team<br>Mechanical Breach/Flash Bang | TL 1 | Target |
| | | | #2 DELIBERATE ASSAULT<br>INTEGRATED MULTI–BREACH POINT<br><br>1 Shooter<br>1 Casualty<br>2 Unknowns | TL 2 and TC | |
| | | | #3 MULTI–BREACH POINTS<br>COMPROMISE<br>2 Suspects<br>2 Hostages<br>Unknown Locations<br>1 Hostage Injured | TL 3 and TC | |

| | | | | | |
|---|---|---|---|---|---|
| | | 1700 | #4 EMERGENCY ASSAULTS<br><br>2 Suspects<br>2 Hostages<br>Unknown Location<br>Medical Emergency - All injured<br>IED | TL 4 and TC | |

*Day 1*

Day 1 begins with the routine administrative tasks and paperwork. I break down the class into tactical teams. Every student gives a short bio about themselves and their team. From this brief description, I obtain a snapshot of their confidence and character and decide who will be my primary TLs. I assign TLs and allow them to select their teams. The teams will work and train together over the next few days and rotate leadership positions as directed.

## SAFETY

True safety begins and ends with the TL. While organizational leaders are responsible for safety, it is the team leader who understands how to manage safety. It is the TL who is going to enforce and monitor safety standards. I show several video clips where teams have failed to train properly. When situations deteriorated, TLs did not step up and take control resulting in catastrophic endings. Most high-level leaders are too far removed to be held accountable as many have not been hands-on for years and are out of touch with current techniques and trends. My goal is to teach TLs tactical solutions to common problems they will encounter, how to avoid pitfalls, and what their job and responsibilities actually entail.

## TACTICAL LEADERSHIP

Under Tactical Leadership, I talk about administrative leadership, training leadership, and combat leadership. Administrative leadership is the boring, but necessary, day-to-day leadership required to keep the organization going. It relates from administrative forms to supply and logistics. Without it, leaders could not sustain the next two levels, training leadership and combat leadership.

Training leadership is just that, the organization, execution, and evaluation of training. This requires administrative leadership as leaders need to coordinate training areas, equipment, time off, role player, etc. Besides ensuring the training goes as planned, many times leaders need to participate in the training to ensure leaders know our duties in a tactical situation.

Finally, there is combat leadership. Training leadership focuses on how to resolve tactical scenarios and how to lead officers safely and efficiently using techniques that mirror real-world incidents (or as close as possible). Simple tactics and techniques coupled with commonsense leadership decisions are the key to success.

## COMBAT MIND-SET OF A TACTICAL TEAM LEADER

I first discuss this concept in chapter 1, but it is just as important during training as it is during an actual fight. Combat mind-set is critical to mission accomplishment. The lack of it is probably the most important factor in most tactical operations gone bad. Either you have it or you do not. Leaders need to continually cultivate combat mind-set and reinforce positive behavior that demonstrates it in a positive light.

Many old-school instructors see combat mind-set as hazing, screaming at, or abusing students while they are performing combat-oriented tasks. In my view, teaching combat mind-set is to teach students how to focus and perform technical tasks under high-stress conditions. Leaders do a good job of praising individuals who demonstrate it in real life, but do a poor job of showing students how to develop and control it.

## LEADERSHIP COUNSELING

As I have stated in the past, leadership counseling is probably one of the most neglected, but important, jobs leaders have. Understand that leaders are in a "people" business and that without proper management of people, our business will not survive.

I point out to students that if they cannot look at someone in the eye and tell them this is where you are at, this is where you need to be, and this is how to get there, you are doing them a disservice. Leaders need to be able to counsel formally or informally as the situation dictates. If you cannot do this, you should not assume the position.

## SELECTION AND TRAINING

Selection and training is almost as problematic as leadership counseling. Many individuals look at the way they were hazed and believe this to be the norm. When setting up and running selection courses, they wish to add to or make them a bit tougher than when they went through them to leave their mark. After time, the entire process becomes distorted and sometimes barbaric.

Instead, evaluate how a student thinks, look at what initiative they took to prepare, their actual preparation work and mental toughness, determine what skill set they bring to the table, and assess whether they are trainable. In reference to trainable, much depends on the student's mind-set.

## PRACTICAL EXERCISES

Most students have never been taught how to run a selection or training exercise. I put the student through a practical exercise and then I explain why I do what I do, what I am trying to assess, and how I measure it as the class proceeds through the exercise.

The first practical exercise I put them through lasts about two and a half hours. I get a snapshot of their overall physical, technical, and inner operability skills. In other words, how they get along under stress. I do this without injuring students.

Students are told what packing list to bring to class to ensure that all students carry the same weight. I assign teams strong TLs for this exercise, those I selected at the beginning of the day. Before we start, I tell the class that TLs will be evaluated by each person on their team. Further, TLs are required to write a counseling statement on the weakest member observed during the exercise.

During the exercise, I keep the next events unknown to add to the mental stress. Normally, I would begin with a brisk walk in their tactical gear to get students warmed up and to assess who is in shape and who is not. This is where I, as the instructor, have a moral obligation to remove students from the class. If a student falls out of the walk, they did not come to the class physically prepared. If I observe this situation and do not remove them for their own safety, I am responsible for injuries occurring later in the exercise. Most of the students I have removed came grossly out of shape and could not walk one to two miles in their tactical gear. Others aggravated old injuries, such as back injuries, and could not continue. This is what I wanted to see to keep them from getting hurt. The problems jump right out at me and I can keep them from hurting themselves.

Next, I check their weapons proficiency with either pistol, rifle, or both. I video record their load, unload, firing, and general weapons handling sequence for viewing the next day. I also save their targets so they can see how they did without argument. I see every kind of weapon-handling problem including no function checks, etc. I also see targets without a round on them after the simple course of fire indicating poor or no zeros. The video review also offers me a nonconfrontational way to critique the class.

In between shooting drills, every team goes on a short run to evaluate their physical fitness levels. In the tactical world, teams must get to the fight, engage in combat, come back from the fight, and be prepared to do it again and again. To give them a pride of ownership, I have TLs lead. I evaluate how TLs manage individuals coming from varied backgrounds, who are unfamiliar with each other.

After the short run, I have students execute another shooting drill, then perform one final physical exercise.

## ASSAULT SEQUENCE

### *Day 2*

Keeping in mind the concept of showing the student the "whole elephant," I review the assault process from planning, rehearsals, deployment, and redeployment to put students from various agencies and skill sets into the same frame of reference. Some may, or may not, have a detailed knowledge or background on the way things should be done.

## ASSAULT PLANNING AND OP ORDER

After the Assault Sequence class, I break down the entire assault planning process and OP "order format." (Simply put, an OP order is "the how" to accomplish a mission or task.) I give students a blank OP "order format," which is about one and a half pages in length. It is simple, generic, and can be used to plan multiple missions. Further, it is not too detailed to cause confusion with the lowest member of the team, which the order should be targeting.

I routinely ask, "Who is the OP order written for?" The commander or the lowest Gumby on the team? My response is the lowest Gumby. Most of the time, critical mission details or tasks are performed by the lowest-ranking person on the team. These people are critical to mission success and must understand the importance of all assigned tasks. I break down the OP order and discuss who is responsible for planning what. I also suggest who can be used to help assist in certain portions as it relates to their job on the ground.

Next, I discuss how the planning process unfolds as it relates to a callout. Everyone has a job, but not everyone will arrive at the same time to the planning area. With that, planning and rehearsals must start ASAP. Teams need to know a generic mission, then they can begin to build a mock-up or outline of the potential target to develop team and organizational flow on the target. Rehearsals, even simple tape drills on the ground, help team members visualize their actions and what their breach point and areas of responsibility look like.

As new intelligence is reported, it must be pushed down to the lowest level. Changes are inevitable and must be addressed. The plan may take a series of changes before the final product is blessed or agreed upon.

## REHEARSALS

Rehearsals are the key to mission success. They can be either educational and productive or just plain painful and a misery to endure. Rehearsals should be run at the individual and team level first, then the organizational level. Teams should be allowed to practice their individual missions first, as team missions are the building blocks to the organization mission. If leaders try it in reverse order, teams will have two problems to contend with: mastering the plan and doing their subunit jobs within the plan.

## COMBAT LEADERHIPS MODULES

- Safety
- Permission
- Emergency Assaults
- Movement
- Breaching
- Multi–Breach Points
- Antifratricide
- CQB
- Medical

- Hostage Handling
- IEDs
- Consolidation and Ex-filtration
- Post Assault Procedures

When I teach the above modules, I try to avoid pushing too many tactics and the "my way of doing business" mentality. Rather, I try and teach how a team leader should best manage the situation and not become too hands-on. For example, if a TL is handcuffing bad folks during room CQB, his focus is too narrow and he is not looking at the big picture of how the overall team is doing. Do they have overall security? Has the room been adequately cleared? Are the cover officers in the best spots?

The same applies to medical emergencies. Too many times leaders want to get hands-on and that is the team's job. TLs need to ensure that they have overall security, security over the unknown patient, someone calls and reports to higher-ups, and someone provides hallway security, etc. If a TL is absorbed in the medical treatment of the patient, they lose the big focus of the overall health and welfare of their team.

Each module has certain protocols or techniques where the leader can best position themselves for the visual, physical, and verbal control of the team. I try and show TLs a few techniques to make their life easier and more efficient.

## TARGET PLANNING BRIEFINGS

During the target planning class earlier in the day, I give each team a "target folder" with a sketch and photos of a real-world target that a team might encounter. I require them to come up with two plans: one using just five personnel in an emergency-assault-type mission, and the other using the four teams of five (as structured in our class). I want to know if they understand the concepts taught in class and if they understand how to efficiently take down the target in a few seconds.

Usually, the ATL briefs the "emergency assault plan" whereas the team leader briefs the deliberate plan. They brief this in front of the class as they would in the real world. Each team has a different target. Teams can generate ideas on how to attack the target should they see the target in the future. I add points and assistance if they get stuck, but for the most part, teams come up with realistic plans. They also determine how many people should be on their teams as they are required to account for all bodies. They learn that if they do not have enough folks, they may have to work with another team in a mutual-aid-type response.

### *Day 3*

Day 3 begins with a safety briefing and inspections, then Deliberate Assaults on targets with live role players and paint weapons. The class goes through the entire planning cycle, rehearsals, target takedown, and debriefings.

I begin with what I call a "confidence target," a simple scenario. I know from experience that the first assault will be rough and it will take a second assault to polish up all the problems encountered in the first run. Putting twenty students from different agencies together for the first time is a challenging task. I must blend movement, breach points, CQB, etc., with chaos added in. I start with only one medical emergency. As the students gain confidence, I start adding more and more scenarios to their plate, to include officers down, IED, etc. I teach them to solve one problem at a time and move on. If every team learns to do this, the entire assault will flow without a hitch.

Beyond tactics, I try and get students to focus on information flows, leadership, and what works and what does not. I teach a primary system and a backup. In short, if their radios die, how do they get the information to higher-ups? With each assault, I assign a new TL for each team so they get experience managing a team under high-stress conditions.

Once the class goes through a series of common assaults, I then begin "emergency assault" scenarios where one team enters first. They are met a few minutes later by follow-on teams. This requires planning and fire discipline, linkup signals, etc. During this phase, they have already practiced the sequence of taking down the entire target. I throw in a twist to make them think a bit more. After each team has had a turn as an emergency assault team, I terminate the training.

## A LITTLE HISTORY AND HOW THE COURSE EVOLVED

When I began this course, I used to lead all the teams on the practical exercise and it became a rite of passage. I made the decision for the team leaders to lead their teams after the initial phase to give them a pride in ownership and to allow them to motivate and evaluate their "new" team. This hands-off approach seems to work and gives the TLs a chance to excel.

During the course I hope to get across the points of "centralized leadership" and "decentralized leadership" and when both are used. Centralized is when a commander controls the entire group up until the point of the assault. Once the command to "go" or "execute" is given, it is now up to the team leaders to drive the pace and tempo of the assault. The commander no longer has control and the success or failure of the mission rests on the shoulders of the TLs. This is "decentralized leadership" and it is up to TLs to push their teams, not lose momentum, and handle any and all situations they are presented. Proper training and rehearsals will make this happen.

## KEY POINTS

- Continue to use video or you will miss a great deal of information.
- Leave your ego out of teaching, especially with new leadership.
- Know when to employ centralized and decentralized leadership.

# 22

# SAFETY

*Getting shot by your own weapon, or another officer's, can kill or maim you as quick as an enemy's weapon.*

—Safety Slide

- **BIG FOUR SAFETY RULES AND A FEW ADDITIONS**
- **OTHER SAFETY VALVES**

## BIG FOUR SAFETY RULES AND A FEW ADDITIONS

This is perhaps one of the most straightforward and critical pieces of information I can impart. There is simply no other way to put it.

---

*Universal Safety Rules*

1. All guns are always loaded.
2. Finger off the trigger until sights are on target.

3. Never let your muzzle cover anything you are not willing to destroy.
4. Know your target's foreground and background.

*CQB Safety Procedures*

5. Weapon is kept on safe with straight trigger finger in a low or high ready until a target is identified.
6. Once a target is engaged, the sector may be swept with the weapon on fire and trigger finger straight.
7. The safety will be engaged prior to any movement or once "clear" is given.
8. Shooter to front has priority of fire.
9. Engage your safety prior to working on injured personnel (hostages/downed officers), to include the manipulation of suspects.
10. Engage safety of downed officer's weapon prior to working on or moving him.
11. Never run behind any target.
12. *When in doubt, don't pull the trigger!*

---

**1. All guns are always loaded.**

Know how to properly load and unload your weapon and do it in a safe direction. I have had individuals discharge their firearm into the pavement while clearing their sidearm in preparation for dry-fire or paint-marking training. Most of time officers had a poor clearing sequence when this happened. I had one individual discharge a round into the cleaning room floor while preparing his pistol for disassembly. He looked at the chamber and did not look at the magazine well. When he let the slide go forward, the round was stripped off, loaded, and then discharged when he pulled the trigger.

Clearing barrels are a great tool to help prevent this or to "capture" the round should individuals not have a good clearing sequence. I know about several individuals who have been shot different times in weapon cleaning rooms because weapons were not cleared properly prior to entering the areas. For those of you in charge of this area, have your aid bags stocked and readily accessible to all.

**2. Finger off the trigger until sights are on target.**

Know the trigger control for the shot required. I use a series of surgical targets for my training on both the flat range and shoot house. I also use other targets to mask threat targets as well as toy "babies," which can be purchased at a local Walmart. I will tape these props in front of threat targets to put added stress on the shooter.

I use surgical targets to find out several things. First, are shooters skillful enough to make the shot? Next, does the shooter understand the line-of-sight and line-of-bore relationship on the rifle in their hands? The typical M4-type rifle has a two-and-a-half-inch sight-to-bore offset at seven yards. Under high stress, many individuals forget and put their sights where they want to hit and forget this offset. Their trigger finger moves faster than their mind. In the end, I would prefer the students not make the shot if their skill set is not there. I instruct them to move to a better firing position.

**3. Never let your muzzle cover anything you are not willing to destroy.**

This rule deals with pistol or rifle muzzles and the direction they are pointed. I teach a "high ready" for the pistol and then a *sul* or "depressed ready" when officers or innocents are in front of the shooter. I teach a low ready with the rifle where the muzzle is below the beltline for better vision.

Rifle issues and violating this rule generally come from students trying to clear through their optics or sights while scanning. By the time they pick up a target, the weapon is already pointed where it should not be pointed.

This rule also applies to slinging a primary weapon. First, I use a common bungee cord that is attached to the front and rear of my vest, on my nonfiring side. With a simple sweep, I can catch it with my nonfiring thumb, pull it out away from my body, and put my rifle through it, securing my weapon muzzle down while I go hands-on with someone, while dealing with explosive charges, negotiating an obstacle, etc. Without a retention device, the weapon will flop around and can be damaged or cause injury. I have seen too many times where an officer pushed the rifle behind him to work on someone only to have the rifle slide around to the front and smack the person the officer was dealing with in the face or head. With the M4-style weapons, the front-sight post connects first and either cuts or takes a chunk of meat out. The older MP5s were much more forgiving when this happened.

One catastrophic incident took place during a shotgun breaching session when the officer failed to put the weapon on safe after breaching a door. When he went to retrieve the shotgun, his finger caught the trigger first and discharged the weapon into the back of his calf with a breaching round. Two rules were violated here. First, the weapon was not placed on safe. Second, the shotgun was pointed at his body.

I heard of other stories of this nature taking place over the years. In the end, secure your weapon where you can control it and keep it on safe.

### *Rifle Reload Position*

Tactical officers and soldiers will reload their primary weapon or clear a malfunction in a tactical environment more often than a pistol. Their battlefields can range from a school hallway to a room or an open field or area.

A technique I have observed students use when reloading the rifle: keep it parallel to the ground or with the muzzle slightly elevated. This works okay on a flat range, but here are the problems I see with this technique.

First, you cannot control where the bullet goes if the rifle accidently fires. The bullet flies straight downrange or up: either way there are no berms in real life to stop this bullet. You lose control of the round. I have seen students take this technique into hallways or into a room scenario, where there are many innocent civilians around the target. Students reload into hostage targets in a room because they trained this way on a flat range. I do not promote loading into your buddies on the battlefield. Students did not take the action a step further and think about people and fellow officers/soldiers downrange. Finally, officers/soldiers did not have situational awareness.

I use load/unload points on my ranges and during training. These points can be simply dirt between two orange cones. In the past year, I have had two instances where rounds were discharged into the dirt during loading sequences. One was due to a firing pin rusted into the forward position due to lack of maintenance and contacted the primer of the round when loading. The other was due to a round sticking in the chamber after the weapon was cleared and the trigger pulled. Both rounds harmlessly struck the dirt in front of the shooter.

A shooter can control where he "dumps" the energy of a bullet and dirt is one of the best natural mediums. A tree also works well. I have even used my helmet or Kevlar vest as a clearing backstop. Launching a round into the community or among your fellow soldiers is not an acceptable alternative. Teach students to be aware of their surroundings and their muzzle whether in a room, hallway, or in the open.

**4. Know your target's foreground and background.**

The shooter is required to make a safe shot leading to the target and is responsible should the bullet leave the target. Slide right/left or up or down to make this happen. I routinely see students new to CQB training hit their point of domination and lock their feet in place like it was concrete. I teach that they can slide left, right, up, or down to change the angle of the shot and tell them where the bullet goes when it exits the bad guy. To take this training to the next level, the instructor should put no-shoot targets in front of the bad target and behind. This forces the shooter to calculate the shot to the target and when it leaves the target. New students will get ramped up to a point where they don't even see the innocent target behind their shoot target. As students get experience, they will see more.

I use an old video of two officers up on a tight porch to illustrate this point. The suspect comes out between the two officers and one officer turns and fires, killing the suspect. The bullet leaves the bad guy's body and hits a fellow officer. In fast-moving gunfights, there will be as many times not to pull a trigger as there are times to make the shot. In short, always look deep.

**5. Weapon is kept on safe with straight trigger finger in a low or high ready until a target is identified.**

Low-ready rifle, high-ready or depressed-ready with pistol; as I have asked before, do you shoot first or do you see first? The answer is that we must always see first. Students trying to make up for a perceived lack of speed think it is faster to go in with a rifle up to a firing position than it is to keep it at a low ready. Blocking your view with your rifle will slow you down and cause you to miss obstacles, furniture, bodies, etc. It will also slow down your discrimination process. You will be looking high and will not be able to see the person's hands down at his waist, where weapons generally are kept. You will see the weapon when it is raised: then your reaction time is the same as his. Only training and discipline will fix this issue.

**6. The safety will be engaged prior to any movement or once "clear" has been given.**

I have a great video of a big-city tactical team performing a live-fire hit after an explosive breach. The point person has an MP5 and appears to stumble coming through the breach point and dumps a couple of rounds into the floor. This has happened more than once and will continue to happen if individuals don't use their weapon's safety.

**7. Once a target is engaged, the sector may be swept with the weapon on fire and trigger finger straight.**

This applies to room CQB and multiple targets. I teach students to remove their finger from the trigger when transitioning targets on a flat range so they will not, under high stress, ride a trigger across a hostage. Sympathetic discharges have occurred while riding the trigger. A shot close to you, a flash bang going off, or an explosive charge can cause someone to sympathetically pull the trigger when riding it.

Soldiers and officers will fall, trip, and stumble during operations. Shithole houses and real buildings are not clean and sterile like flat ranges. I teach how to come off the trigger while shooting on flat ranges because I expect them to do it in a high-stress tactical situation.

**8. Shooter in front has priority of fire.**

I routinely observe shooters shooting over the top of their teammates or fellow students from several feet back both in training and in short combat video clips. This is dangerous. Four out of five times the shooter will get away with it, the fifth the shooter will shoot the guy in front of him. You have a hard focus on the target and the guy downrange/in front of you may decide to move left/right up or down to get out of the line of fire or make a more effective shot. When the shooter does, the shooter will move into the path of your bullet(s) before you can react and come off target. I have seen one department violate this rule twice in two years, each time shooting and crippling an officer to the front. I see many instructors in live-fire video and in still pictures violating this rule with their students. They are lucky for the moment.

**9. Engage your safety prior to working on injured personnel (hostages/downed officers), to include the manipulation of suspects.**

Your adrenaline will be pumping and you do not want to accidentally shoot a patient or a suspect you are controlling. This has happened before. Engage the safety of a downed officer's weapon prior to working on or moving them. Some policies suggest that you disarm unconscious officers as they may come back into the fight when they awaken.

This rule overlaps with the muzzle awareness rule. Your weapon should be on safe and stowed while you are treating the patient. In medical scenarios, officers have plugged their flash suppressor with sand from kneeling down on the floor. This small amount of sand will cause you to lose the first round fired from your weapon at any distance. The shooter is creating an "angle of deflection."

I observed officers take a steady rest and try and engage a six-by-thirteen steel chest plate at forty yards and miss because of the sand. It blows out after the first round fired, but you will miss the first shot fired. This is a reason why I use muzzle caps and keep a few spare on my body. They are made to be shot off the weapon and not change your zero. I have tested muzzle caps and they work as advertised.

It is your job to create a safe environment to treat the injured. This may mean moving them to cover first. If their weapon accidentally discharges while dragging it, it might cause injury to them or another officer/good guy. The same applies to your weapons and their posture/safe condition.

**10. Engage the safety of a downed officer's weapons prior to working on them or moving them.**

See above.

**11. Never run behind any target.**

Whether in a shoot house or on a field range, knock down targets prior to passing them. If you do not, someone may see and shoot the target and not see you behind or on the other side of it. This has happened many times in my military career. In the shoot house, I use targets that can be knocked down when executing immediate threat drills or when moving to points of domination through people (targets). You may discriminate well. Your partner may not. In real life, a gun may come out after you have passed by someone. If they are that close to me, I will generally push them off into the kill zone.

This also applies to a field environment when soldiers are doing fire and maneuver drills. I knew of one case where the two fire teams were separated by a small clump of bushes. One team bounded forward and one of the soldiers moved behind a target. One soldier on the other team did not see him move and fired on the target. One round struck the downrange soldier on the left side of his face and came out his nose, crushing his cheekbone. Another inch and it would have been a fatality.

**12. When in doubt, don't pull the trigger.**

All students come in with a different skill set and experience level. The instructor must train how to react and to look and interpret various situations and act accordingly. This takes time, patience, and experience.

To add, law enforcement and civilian tactics training may require two skill sets. When students take a knee to engage, they should be taught to scan behind with their head and not a weapon to ensure no one is shooting close to them. The police will be arriving soon. In the civilian situation, you are an unknown to the officers, standing over a body with a weapon. Officers will react to what they see.

The civilian shooter just fired several rounds, has auditory exclusion (hearing is impaired), and may not hear police officers shouting commands. If you turn with a weapon exposed, there is a good chance you are going to get shot. Responding officers only know what they see, an unknown person standing over a body with a gun who is turning toward them with a gun. When law enforcement or military personnel turn, they are in a uniform and can easily be distinguished from the bad guys.

Students who grow up in one discipline only know what they have been shown. Their instructors may have a limited skill set and limited experience and will not find out why something is dangerous until someone gets shot.

As for some of the comments from armchair commandos, I will leave you with this. Unless you have been in Special Ops for a few successful years, you don't know squat about what commandos do or do not do. I read articles and books all the time where an individual will reference a short video clip of soldiers shooting in training or combat actions and say, "See, they are doing it." Yes, they are. And they are probably doing it wrong or in an unsafe manner, especially shooting from behind or over individuals from an unsafe distance. I see it all the time. This includes some Special Ops guys.

The military does not always have the best training available, and when screwups happen, they are mostly covered up, or in combat, attributed to enemy fire. The military does a poor job of passing down lessons learned and many times spends more effort training foreigners than their own deployed troops.

Further, as the current conflicts wind down, there will be more instructors coming to the United States to teach and instruct. Check out their credentials and ask about their safety protocols. If the credentials are weak or shallow, be suspicious. I would be suspicious of an instructor who was in a Special Operations unit for only one year, who then leaves or is pushed out, and decides to be a civilian instructor. Did he get kicked out for screwing up? Is this the person I want to model my training, life, or profession after? Probably not.

Finally, if I could not train students safely, I would be out of a job. This is why I teach safety as my first block of instruction in all my tactical classes and continually hammer it home throughout the training. I have a successful system that has been developed during thirty years of training, refining, and reviewing why accidents happen. I try and learn something each day I train and still consistently adhere to my safety protocols.

As for the statement that safety is unattainable, this is false. This is only true if your ego is too big or your experience level too limited.

### *Weapons as Tools for Blunt Impact*

I have used the rifle as an impact, deflecting, and pressure point tool. Many trainers and chiefs are repulsed at the thought of striking someone with a weapon, but it is the logical next safest step when using force. Weapon impact tactics provide another choice in the use-of-force spectrum besides deadly force.

First, striking someone with a pistol is not generally advisable. If you strike or jam (punch) someone with a Glock, you are most likely going to induce a weapon malfunction. The 1911 style pistol can be used to punch because the safety in the "on" position keeps the slide locked and the weapon intact.

Rifles can be used to strike individuals with light, medium, or hard blows to either soft tissue or bones similar to that of baton training. You can also deflect and move people out of your way with your rifle. Finally, I have used it as a pressure point device to gain control of a subject without taking my hands off the weapon.

The only other option when putting your hands on people in close quarters is to take your hands off your rifle during movement to deal with someone. This means you are manipulating someone with a hot weapon between you and him. Not a safe or good idea. Simple and effective moves can be practiced and integrated into department defensive tactics programs and should be taught. It is another tool that you can safely use before you escalate to lethal force.

## OTHER SAFETY VALVES

### *Electronic Ears*

I religiously use electronic ears when instructing to help ensure all possible safety concerns are identified in a rapid manner. While shooting demonstration drills, I use soft ears to attain proper head-eye alignment; all other times I use electronic ears, which can be bought for under $50. They allow me to hear all the problems on the line (student concerns, bolts being manipulated, etc.) while I am giving firing commands. During CQB training, I can hear when a student "rolls" his rifle safety to fire while entering a door. This is a bad habit I try and break early on in training. Students should only place their weapon on fire when they have identified a threat target.

### *Portable Public Address System*

When training more than three or four students, I use a portable public address system to relay fire commands. Many students do not own electronic ears, and the inability to hear range commands is a definite safety concern. Seeing a safety violation and stopping an action in a rapid manner is also crucial to running a safe line. In addition to safety, a portable PA will save your voice over a multiple-day shooting class.

### *Medical Kit*

I keep a basic medical kit on my body as an instructor and also keep an aid bag centrally located on the range. My basic medical kit consists of the following:

- Tourniquet (1 on body and 1 attached to my rifle stock)
- Medical scissors
- QuickClot Combat Gauze
- Kerlix - 2
- Co-Flex Wrap - 2

I recommend placing this basic issue on your body, in a spot where you can reach it and treat yourself, if needed. If you cannot reach it, you cannot treat yourself should the need arise. This is also a good fit for an "Active Shooter" bag. I like the idea of immediately being able to treat and stop bleeding without having to run one hundred yards to get the range aid bag.

A dedicated aid bag for the range should have the same items as listed above and a few more. Band-Aids are good to have along with a collapsible litter. Band-Aids are for your common nicks, cuts, and gouges. Collapsible litters are for moving injured or unconscious people for any distance.

### *Paint Pens*

I use a common paint pen that you can buy at Walmart or a discount store to mark sight zero position and to mark the off side of the M4 weapon safety. I draw a straight line across the safety onto the receiver of the weapon. A quick visual scan can alert me to a student running a weapon on fire or moving in this condition.

## KEY POINT

- Everything listed. Reread this chapter.

# 23

# TECHNICAL SKILLS, TACTICAL SKILLS, AND THEIR INTEGRATION

*Our tools and how we use them is what separates professionals and mediocre performers.*

- **TECHNICAL SKILLS AND TACTICAL SKILLS**
- **FLASH BANGS OR NFDDs**

## TECHNICAL SKILLS AND TACTICAL SKILLS

Technical training is the skillful way you employ your weapon systems. Tactical training is the thought process and action you put into that employment. Physical is just that, a condition that allows for sustained tactical and technical engagements with one or more threats in an efficient manner with rifle, pistol, or hands, over and over again on demand. I classify technical skills as the following:

- Flash Bangs or NFDD
- Shotgun Breaching
- Explosive Breaching
- Medical
- Hand Combat
- Less Lethal

## FLASH BANGS OR NFDDS

These skills require specialized technical and tactical training for their integration. For example, NFDDs/flash bangs are an eight-hour block of instruction that will "certify" the end user to safely employ the device. The class consists of several hours of lecture accompanied by a PowerPoint presentation. Next, the students dry practice employment with the use of inert training or downloaded munitions, and finally students will employ full-strength munitions in either an open or enclosed environment.

### *Tactical Integration*

The biggest training gap is where the technical instructors do not make the leap to the tactical side of employment. Many times instructors are "technical" guys and feel comfortable teaching the classroom portion of the material and not so comfortable with the actual employment. The operators need the tactical integration, meaning how to take home the skill and integrate it into their team. This goes as far as what pouches to use, where to set them up, and how to best employ the device weak hand or strong hand. Instructors who rely on the flash bang to "awe" the student with its big boom are not maximizing the training time.

Instructors should show various types of tactical employment and their pros and cons: that being strong hand weak hand, on call or preplanned, right deployment or left deployment, window deployment or in conjunction with bang poles. After a sterile throw of their first "banger," students should be broken down into teams to move through the target employing the NFDDs on open doors, closed doors, doors that have to be breached, and no-bang scenarios where there is cause not to deploy the device.

Once the student reports back to the team, the team leader needs to continue reinforcing the training of the individual and reviewing the team's SOPs on deployment, to include carry, hand and arm signals for deployments, automatic deployment situations, or ones on call. Shotgun Breaching, Explosive Breaching, and Manual/Mechanical Breaching should all be taught in a similar fashion. Listed below are a few training contingencies that a TL should address in the technical modules and tactical employment:

---

## NFDDs

- Right throw
- Left throw
- Dud
- Bounce back
- No breach
- Ride the banger/initially pin from doorway

BREACHING

*SHOTGUN BREACHING*

- Successful breach
- Delayed breach
- No breach
- No breach with mechanical backup
- Second gun up
- Immediate threat on breach

*EXPLOSIVE BREACHING*

- Priming procedures
- Tactical charge emplacement
- Successful breach
- Compromise prior to breach
- Compromise on breach
- Misfire procedures
- Dual breach setups

MEDICAL

- Patient down in the open
- Patient down in the room
- Patient down in a hallway
- Self-aid
- Buddy aid
- Unknown-subject aid
- Suspect aid

HAND COMBAT

- Control techniques
- Handcuffing both hard cuffs and flex ties
- Offensive hand combat
- Knife techniques
- Pistol retention and offensive movement techniques
- Rifle retention and offensive movement techniques

LESS LETHAL

- Less lethal with lethal coverage
- Immediate compliance
- Slow compliance
- No effect/compliance
- Lethal threat presents itself

In the end, the key to maximizing technical and tactical training is up to the instructor. I believe it is a waste of time, money, and resources not to integrate tactical training into technical modules when time and resources permit. Most, if not all, technical classes have a direct tactical application, and this should be addressed and integrated into the training.

In the end, it is up to the TLs to determine what technique they will use to employ the technical device or skill.

## KEY POINTS

- When teaching technical skills, always integrate tactical skills and applications into the training.
- Use training sites and venues that replicate real-world scenarios when executing this training.

# 24

# MAINTAINING THE INSTRUCTOR EDGE

*Use the butane tank for cover.*

—A Tactical Commander

*Perception is reality.*

—A true but most annoying saying

- **PROFESSIONAL TRAINING**
- **WRITING ARTICLES**
- **WRITING BOOKS**
- **ATTENDING MAINTENANCE TRAINING**
- **SEMINARS AND CONFERENCES**
- **REVERSE ENGINEERING TACTICAL SCENARIOS**

## PROFESSIONAL TRAINING

As mentioned earlier, there are both pros and cons that come from training with multiple instructors. Working with many different instructors can raise your overall awareness of techniques and tactics, but at one point you are going to have to settle down and find one system that fits your physical make up, job, etc.

In my own personal experiences, I was doing fine with one system of shooting, then another instructor would come in and show me his system. Frustration would set in and I would have to take a step or two back to make any progress. I would start to ask myself, why is this instructor teaching this technique? Is it for real world or shooting games? I would not accept the technique until I answered these questions.

I made a decision midway in my Special Operations career to tailor my personal and professional system to combat and not IPSC or fun shooting. IPSC shooting techniques will help you be a better shooter, but they will not make you a better tactical shooter and can actually create bad habits. I know of one instance where a shooter died from using speed versus tactical shooting technique while responding to a real-world active shooter incident.

## WRITING ARTICLES

Start small. Discuss something you are passionate about, know something about, and that is relevant. Once you write an article, have someone you know read it for content, punctuation, and grammar. Writing helps you develop logical and systematic thoughts and how to relay those thoughts to your readers. Articles also establish your credibility. If you design your article series correctly, you can make them into a book.

When writing articles, try not to personally attack, but rather, attempt to illuminate the subject with positives on what can be done and why. Show more positives than negatives and use pictures, studies, and graphs to illustrate your points. I do this with rifle ballistics. Everyone and their brother want to throw a graph at readers in reference to ballistics. I show pictures of my targets and my sight pictures with the bullet holes highlighted. They do not lie; no one can argue with them.

Finally, articles are some of the best forms of free advertising. Generally, I take my finished articles and put them on my website as a resource for anyone who has the need for them. One of my most popular articles was about the use of the weapon safety. I still get requests to reprint it. When I started the business, I wasted a great deal of money buying ads in magazines which produced little or no business.

## WRITING BOOKS

When you have enough to say and want to say it positively, write a book. As a student, you are going to be exposed to a great deal of negative training in your life. Remember why it was negative and articulate why it did not work and give positive solutions on how to fix it.

## ATTENDING MAINTENANCE TRAINING

I like to attend training from time to time as a student. It is difficult to find an instructor that will let me do this. If I do not know the instructors, they may think I am trying to steal their information. Others may find it uncomfortable when students find out who I am and start asking me questions instead of the primary instructor. I think it is good to continue my education if it can be done in a controlled and tactful manner.

I get the instructor's permission to use my own tactics and techniques. Otherwise, I will be setting a bad example for other students. My ideal class would be to shoot my own system during another instructor's course and be able to pick and choose what I take back. Further, I would be able to see if my system works with the drills I deem useful.

## SEMINARS AND CONFERENCES

Generally, attending seminars and conferences allows me to provide a snapshot of my training to a variety of individuals from many agencies. This approach helps me establish credibility in the field with all costs picked up by the host organization. I get the chance to give back, show my system, and touch a great deal of people with a one-on-one meeting. It builds rapport.

Seminars can be short or long, one-hour blocks to all day long. Generally, I teach an eight-hour block for my Leadership and Training for the Fight seminar. I conduct a longer block in the morning when individuals are rested and shorter blocks in the afternoon with frequent breaks. Handouts help keep students in sync with the slides, and I have received positive feedback on my handouts. I have found that I constantly review, refine, update, and polish my seminars. I add relevant material and delete redundant or outdated items. This keeps the information that I present fresh and relevant.

## REVERSE ENGINEERING TACTICAL SCENARIOS

As my career as an instructor evolves, I constantly read about officers and security personnel being killed or injured on duty while alone, in groups, in cars, etc. As an instructor, I pride myself on looking at these incidents and ensuring my training methods address these issues. If not, then I work to develop and add to my basic instructor drills and curriculum.

The first major training void that comes to mind is "officer down" scenarios, where the patrol officer is ambushed, bleeding, and down under direct fire of the suspect. These training additions require both a tactical solution and an administrative agreement for the use of force.

An "officer down" scenario is the same as a "hostage rescue" scenario. If the hostage was a civilian, officers would attempt to recover them in a rapid manner. This same logic should apply to police officers. The scenario can be handled in the following manner:

A. Let the officer bleed until the suspect surrenders.
B. Recover the officer and be prepared to put direct fire on the suspect.
C. Attack the suspect with tactics similar to an "active shooter" response while recovering the downed officer.

Both B and C are correct responses in my book. The organizational problem is that I have to plan for this contingency and then ensure my chain of command supports my tactics in training and a real encounter.

The tactical solution must be found in developing common worst-case scenarios for this situation. The officer is down either in front of a breach point or in the open. Tactics must be developed for both positions, practiced repeatedly, and videoed to refine for SOPs.

As an instructor, I can be proactive and define or identify problems and help find solutions, or I can be happy with the status quo. When I first started teaching law enforcement officers over ten years ago, many of the tactics and aggressiveness I teach now would not have been considered an option. Times change, as does the level of violence in our culture. Since I started instructing, the Columbine massacre took place, 9/11 happened, Mumbai was bombed, and four police officers were murdered in Washington State. With this escalation of violence, officers must increase their level and intensity of training and refine their tactics.

## KEY POINTS

- Maintain your skills and personal training.
- Write articles and continue to refine your mind, teaching, and tactics.
- Remember the words of John Cotton Dana: "He who dares to teach must never cease to learn."

# 25

# SELECTION OR TRAINING COURSES

*Selection is a never-ending process.*

- SHORT-TERM SELECTION PROCESSES
- SELECTION, TRAINING COURSES, OR BOTH
- PRESELECTION
- TARGETING GROUP OR INDIVIDUAL
- SETTING UP THE COURSE
- EVALUATIONS—WHAT TO TEST AND WHAT TO TRAIN
- TWO-DAY SELECTION
- THREE-HOUR SELECTION
- OVERALL TRAINING
- ORAL BOARDS

## SHORT-TERM SELECTION PROCESSES

Selection is one of those all-important topics that must always be at the forefront of any tactical TL's thought process. I address this topic in part I of this book during my discussion on the qualities that define effective leadership. I bring it up again here because the selection process is critical to successful training. Watching the progress of a student can provide critical information on that individual's capabilities as well as his physical and mental stamina. This data can greatly affect what positions that student is placed in after training has ended. Similarly, that data can dictate what roles a student fulfills during the training session itself. I cannot say it enough: selection is a never-ending process.

During my time in Special Operations, the command conducted an occasional gut check to evaluate the level of performance and readiness of the men. Physical training was done on our own and we were expected to maintain the appropriate level of fitness. Shooting events were sometimes thrown in to ensure that we were keeping our marksmanship skills up.

Striving to use the same system, I modified the gut checks to fit both ROTC and law enforcement programs. Gut checks provided a "snapshot" of the individuals and their preparedness. They also showed which teams and TLs were doing their jobs (for more details on ROTC gut checks, see chapter 3).

### *Law Enforcement Gut Checks*

I modified my gut checks for my Law Enforcement Tactical Team Leader classes by adding technical skills, such as rifle and pistol marksmanship. I had my students perform the initial walk in all their gear, then shoot their long gun wearing gear during a general course of fire. We would execute another physical event followed by a pistol course of fire. I would video the gut checks and play the video the next morning during the debriefing. I also hung their targets around the classroom for them to see.

During certain portions of the event, I put TLs in charge of their newly assigned teams. When the gut check was complete, the TL started a written evaluation on their weakest team member and gave it to me the next day. Also, each team member gave me a written evaluation on the TL's performance.

## SELECTION, TRAINING COURSES, OR BOTH

If you only had three hours, two days, or three weeks to perform a training and selection process, how would you conduct it?

I recently gave a lecture to a group of federal law enforcement officers who administered a selection process in what I term "old school" techniques. That is, they used physical and mental stress and punishment. The law enforcement group mentioned that they were not getting the numbers they wanted and were losing two-thirds of the candidates in the class. Through a series of direct questions, I wanted to find out if the law enforcement group got along with their overall organization and how candidates who were not selected viewed the selection group after the candidates returned to their offices. The selection cadre admitted having problems with regions and the leaders. Sometimes candidates that they deselected or terminated harbor grudges when they move into leadership positions in regions and cause future working relationships to be strained. I thought about their dilemma, and here is what I suggested:

*First, run the selection course as a "training course" for everyday officers in their unit. In this unit's case, they have three to four weeks to conduct this selection process. My thought is simple: run a nonharassment selection process and actually train all officers in an accredited course, then select or offer jobs to the ones who have performed well and met the criteria.*

Why no harassment? Exceptional training courses can be run with no harassment. Students will put more pressure on themselves to perform than any instructor can generate. Also, students will learn and retain more without harassment. Many instructors will argue that you need the stress and harassment to weed out the weak and those who are not fit for the job. I would argue this point. Most selection instructors who harass only know this: I was harassed, and by god, I am going to harass someone if I get the chance. They have been poorly trained and do not see the harm this type of training inflicts.

Further, most individuals in today's society are coming out of academies with less stressful training than those of years past. Asking today's rookies to try out for a tactical team means a great leap from their watered-down academy training. Academies are pressured to produce and are striving for numbers. The organizations do not want to lose too many cadets. One concept they subscribe to is this: Let patrol or the field training officer (FTO) make them or break them. Then, FTOs are pressured to get rookies to the street to let the street make or break them. This is a poor way of doing business.

Many times, individuals want to perform better and serve in higher positions, but do not know how to get there. I believe that we should give them that opportunity. With the gap created between operational units and the academy, the operational units need to take on the task of training the individual to the required level.

## PRESELECTION

When an announcement is made for a selection/training class, physical and technical goals should be addressed and entry standards specified. It gets expensive flying people down to a selection class who are only sent home because they fail or quit because of their inadequate physical condition.

Let's face it; you have to be in basic shape to wear tactical gear for a week and train. If you are not in shape, you cannot possibly train. You will only think about how heavy all this equipment is and how much it sucks instead of absorbing the information presented. I suggest that candidates submit to a physical training test before they leave. If they fail, they do not get a ticket to show up. This will save a great deal of money in the end game.

If you do not want candidates to form bad habits in a technical skill, do not require them to practice the skill before they come to the course. For example, if you are going to train students in shooting, do not have them practice their own style of shooting. They will probably practice it wrong. There may not be enough time in the course to untrain and retrain the student. I had this happen several times during my civilian training and it is stressful on the student to "unlearn" bad habits while under time constraints of a week-long course. As much as students wish to get ahead of the class, they must understand the reason they are taking the class is to learn the new techniques.

## TARGETING GROUP OR INDIVIDUAL

Structure your course to test the individual, the group, or both. The problem with students initially going through a section as a large group, or element, is that this allows the weaker individuals to blend in or hide. The group has the ability to carry these students along. Targeting individual drills gives a better snapshot of the student. In a three-week course, the first two weeks can be individual assessments and the third week a group evaluation to include leadership.

## SETTING UP THE COURSE

I suggest setting the course up as a quality stair-stepped training course that puts the academy to shame in respect to professionalism and instruction. The goal should be to hear students mutter upon graduation that it was the best training that they have ever received and that they hated to go home.

I suggested the following to one agency:

**Week #1**

Rifle Marksmanship
Pistol Marksmanship
CrossFit Instructor Certification

**Week #2**

Rifle Marksmanship
Pistol Marksmanship
Basic CQB
Medical
Flash Bangs

**Week #3**

Comprehensive Tactical Course such as Raids/Advanced Hostage Rescue
Exterior Movement
CQB

Flash Bangs
Interior Movement
Raid/Hostage Rescue Planning
Scenarios

Week 3 ties weeks 1 and 2 together. Promising individuals can be plugged into leadership positions for additional evaluations. Give two certificates to the course: a CrossFit Instructor certification and a three-week Raid or Advanced Hostage Rescue Course certificate. What do we accomplish with this model?

- Save a great deal of money.
- Better training for other officers from the same organization and as successful students return to work better qualified to perform their job.
- When these students return, they praise the training and cadre instead of spreading hate and discontent. These future leaders in the organization will be more apt to send their subordinates to the course.
- You have given back to the overall organization in training and money.
- You set a new training standard for the organization and it puts pressure on the academy to emulate your training and behavior.
- You build rapport within your organization and area leaders are more apt to work with you and help you should you need it.
- You set the example for everyone in how to properly train.

Week 4 is reserved for formal interview boards, team assignments, and TL integration to the force. Simply send back all individuals not selected on week 3 and hold over the potential candidates.

## EVALUATIONS—WHAT TO TEST AND WHAT TO TRAIN

Sit down and list with your cadre what you are looking for in an individual. You should assign a cadre member to counsel each team and to keep track of individual performance. Other cadre members can report their observations to counselors on a daily basis.

During my time in ROTC, I modified "blue cards" and used them in gut checks and counseling. These cards list almost every leadership trait and value you can think of. The scoring system is simple, E – excellent, S – satisfactory or N – needs improvement.

I recommend creating your own cards based on this simple and efficient grading system. On the front of a card, you write a simple narrative using the STARR format:

S – Situation

T – Task

A – Action

R – Result

R – Recommendation

On the back of the card, you simply check the appropriate boxes.

### *STARR Format*

*Situation* refers to an activity or task. I include a date and time and setting for the event. For example, "Officer Smith was conducting active shooter training as part of an annual department requirement on 25 November 2008 at about 0900 hours."

*Task* refers to the job or specific task. For example, "Officer Smith was moving as part of a two-person element down a hallway during active shooter training paint-marking scenarios involving live role players."

*Action* refers to what they specifically performed, good or bad. For example, "Officer Smith lost physical and emotional control, slumping to the floor against the wall when fired upon, turning his back to the offender screaming, 'Oh shit.' Officer Smith remained dysfunctional for several minutes while a fellow officer was hit and losing blood during simulated role play."

*Results* refers to what happened, good or bad. For example, "Officer Smith quit during the fight, mentally gave up, and left his partner to neutralize the bad active shooter role player by himself in a one-on-one gunfight. Further, Officer Smith failed to respond to fellow officers' need for medical attention when requested."

*Recommendation* is what the instructor recommends to improve student performance, better their actions or thought processes, or to praise good. For example, "Officer Smith, you should immediately attend remedial training or another Active Shooter Training class to improve your performance. If you cannot improve your performance and maintain physical and mental control, you should seek employment elsewhere."

Think long and hard about what to test and what you want to train. If students show repeated behavior that is unacceptable, do not be afraid to document it and counsel them. You owe it to them as a leader.

As for physical training, you will not get them into any type of shape in one, two, or three weeks. Why not train on how to train, such as a CrossFit Instructor course, and allow them to take skills back to train themselves and their peers? Again, you will not get anyone in shape in three weeks, and trying to do so can cause injury.

Also, if you are going to teach shooting, it is easier to start with a blank slate than to try and erase ingrained habits. As long as they are safe, we can make them better.

## TWO-DAY SELECTION

In the law enforcement arena, I would conduct a two-day active shooter training course to select potential candidates. The reason for this is simple:

- "Active shooter" training is necessary and required. It needs to be performed in any case.
- "Active shooter" skills are based on "hostage rescue" techniques and you are already training potential candidates in future skills.
- You can assess all the individual technical, tactical, mental, and physical skills during this training.
- If you learn to teach a class, you will know the information better as an instructor and better polish out your personal teaching skills.

Using "active shooter" training as a selection process, you give back to the department and are efficient with your time. You also are working and getting to know other department personnel and creating a better work environment versus an "us and them" relationship that is common in most departments between officers and tactical teams. Further, you have one system of training instead of two courses that do not complement each other during tactical operations.

## THREE-HOUR SELECTION

The fastest way of accomplishing a short-term selection and assessment is the gut check described earlier in this chapter. By telling candidates ahead of time what they will be tested in, you will find out a critical attribute which is initiative. I would advise a law enforcement officer who is going to a selection and assessment that involved shooting skills to go to the range and ask the range master to work with him, or ask a better shooter to help develop his skills. He should go to the gym to work on strength and pound the pavement to develop cardio endurance. This system does not teach as much as the two-day and three-week modules do, but you may have time to rerun certain scenarios to teach candidates how to respond in a proper manner. Finally, you would be better served to give each participating individual a debrief or after-action report of his or her performance and see if he takes those critiques to heart, improves his weaknesses, and comes back to the next selection process.

## OVERALL TRAINING

The training course outlined in the three-week module is a rough draft of one that covers many of the common deficiencies in the training world today. As in selection, training is a never-ending process. Look at your organization and their mission, list their critical skills, and put as many of them as you can into selection/training programs.

## ORAL BOARDS

Oral boards can be conducted passive, neutral, and aggressive. I have participated in all three. Some individual board members can be instructed to be the agitator of the group. Others ask pointed or shocking questions. This needs to be uniform and practiced ahead of time to ensure that each candidate gets the same treatment. Again, there should be a point in pushing buttons, such as the hot temper of a given student.

These boards can be accomplished while the candidate is fatigued, such as immediately after a physical event, later that day, or the next day.

## KEY POINTS

- Look at developing a harassment-free program as discussed in this chapter.
- Give back a better student than you received.
- Remember: train before you test.

# 26

# TRAINING COMBAT MIND-SET

*Most of the time you cannot choose the time and place of the fight, but you can choose the outcome with your technical, tactical, and physical training, as well as your combat mind-set.*

- **COMBAT MIND-SET—GENERAL**
- **TRAINING TO THINK WITH A COMBAT MIND-SET**
- **COMBAT MIND-SET IN TRAINING**

## COMBAT MIND-SET—GENERAL

I debated on whether to write on combat mind-set and the ability to infuse it into students. Some folks say that you are born with it, some say you have it, or you do not. Others believe it can be cultivated.

I am in the group that says it can be cultivated in most individuals. I believe it can be cultivated in those who wish to attain it. There are those individuals that are CGEs (certified grass eaters) and either do not believe in getting involved in the protection of others, or place more value on their own skin, or simply believe that violence accomplishes nothing.

The key to teaching and instilling combat mind-set is to reach and strike a chord that lies within every person. That chord is the belief in the right to life, freedom, and happiness and the duty to protect others they care about. An individual has to make a choice as to whether these values are worth fighting for. Instructors teach how to find peace of mind with this decision. The rest is easy.

## TRAINING TO THINK WITH A COMBAT MIND-SET

Over my lifetime, I have seen a multitude of techniques, both good and bad, on how to teach and infuse combat mind-set into individuals, soldiers, and law enforcement personnel. Several factors come into play:

- Fear
- Anger
- Emotional control
- Coming to terms with life and death

### *Fear*

Fear is a mental state and the most common emotion. It is simply allowing uncertainty to fill your mind rather than filling your mind with positive solutions to the problem. To help eliminate uncertainty, you must put the problem in context.

When you live in fear, you are the type of person who is reactive and looks at what you cannot do, instead of what you can do, to make a positive impact. You clutter your mind with useless information. To put it simply, if you allow your mind to be filled with doubt, then that is all you will know. You will be unable to focus on possible solutions.

A case in point can be seen in the time when I was a high-altitude–low-opening (HALO) military free-fall jumper and jump master. For the record, I hated jumping and did it because it was a job requirement. I was not happy about jumping from a perfectly good airplane and saving my life with a piece of silk. Something about the gravity equation. My mind would calculate all the problems that could come into play. I imagined all sorts of possible malfunctions with the parachute, entanglements with other jumpers, plane crashes, etc.

In basic Military Free Fall, they taught enough to survive. I was sent to the MFF (Military Free Fall Jumpmaster) course where I learned a great deal more about the mission. I learned the technical aspects of the gear, how to inspect it, how to spot, calculate wind drift, etc. This knowledge gave me a great deal more confidence and more things to think about. I replaced the doubt in my mind with positive thought and action. The additional action and thought was to care for the jumpers under my control.

### *Anger*

Some individuals see life clearly and are irritated by ignorant and self-serving people, and are swelled with anger. These individuals either show it, conceal it, or do both. These folks are the most easy to channel and develop a positive combat mind-set as they simply have to redirect their emotions.

They can clearly see the problems and the solutions to the problems. They become frustrated with people who are not aware and are self-centered/self-serving and prefer to live in a vacuum with no compassion for people around them.

### *Emotional Control*

I have talked about being "high speed" in part I of this book. Besides bringing all the skills and tactics to bear under high stress, high speed is also about being emotionally neutral and calculated.

This mind-set, or thought process, needs to be practiced in high-stress or high-risk training to help perfect it. Remembering Somalia and moving to the first crash site, I was continually calculating the temporary situation, potential danger areas, the location of my team, next areas to move and control. I would also think about the enemies, their capabilities and marksmanship ability, angle of incoming fire, and what they could see and not see. I did not think too far ahead, but tried to solve one problem at a time as the situation changed and developed. When approaching and clearing corners, I knew that only two possible things could happen. Someone was either immediately around the corner or farther down the street. When I took the corner, I kept this simple fact in mind. I am going to do either A or B. Not much else.

### *Coming to Terms with Life and Death*

To find peace and direction, you need to come to terms with life and death and understand what control you have and do not have in life. Yes, you can take a round in the forehead from an untrained youth, shooting an unzeroed AK-47 two blocks away, who knows nothing about sight picture or trigger control. Most likely, this will not happen. There are little things you can do to greatly enhance your survival into old age. You can take care of yourself physically and stay in shape. You can be aware of your surroundings and situation at all times; as backup, have simple plans for problems that may arise.

### *Channeling Thought*

It is said that those who are more aware may also be more pressured by fear as they are alert to it. Fear is uncertainty. Uncertainty is negative thought. This must be replaced by productive and positive thought. Sounds simple. It actually is. It just needs to be practiced and reinforced.

Teach how not to let fear creep in and how to channelize actions into successful effort. Leave no room for fear. If we fill our minds with productive thought and proactive action, we will not have time to fear. We will simply go from positive solution to positive solution, attempting to think about and implement positive and decisive action.

## COMBAT MIND-SET IN TRAINING

Be quiet, confident, positive, and informative as an instructor. Prepare students mentally for what they are going to see, both in your lecture and in your scenarios. In your lecture, use relevant stories, photos, and video clips. Go into detail why your tactics and techniques are important.

When you begin your fieldwork, replicate your scenarios to real world as close as possible. Use blood and torn and tattered clothes on victims. When possible, have officers cut those clothes off. Train as close to the edge as possible. Have role players rehearse several courses of action: surrender, initiating the fight, dying quickly, and being shot to pieces. The more scenarios officers experience in training, the less hesitation they will have on the street. You are preparing them mentally and physically to win.

## KEY POINTS

- Train students to act.
- Once students are trained, instructors should throw surprise scenarios in from time to time.
- Make scenarios as realistic as possible.

# 27

# THE PROFESSION OF TRAINING

- **PROFESSION OR BUSINESS**
- **HISTORY**
- **ETHICS**
- **SAFETY**

## PROFESSION OR BUSINESS

With the winding down of the conflict in Iraq, we may find ourselves with a wave of new instructors in the business. The result will be mixed. Few will survive on their own. Many will be gobbled up by existing companies. Others will band together to form small and flexible companies that teach and provide security services to ensure their abilities net them as much profit as possible.

This type of business can be good, bad, or mediocre. Generally, they attempt to get enough contracts to survive. Some will be what I term one-hit wonders—individuals who will eventually deliver their product, and then realize that they will not get a call back for future training due to the quality of their classes.

Training dollars are always in short supply and law enforcement agencies are always looking for the most bang for their buck.

Law enforcement and military units should also be looking for continuity. That is, do all the training company systems dovetail into and support each other? If they do not, agencies may be taking a step back before they take a step forward. It may also cause havoc in their current training system.

Systems professed as being on the "cutting edge" are many times not. They are retreaded systems that have been taught somewhere else and their origins and reasons for use have been long lost.

The problems lie in the fact that these new trainers are scrambling. They made high dollars overseas and are now working at U.S. income levels with a much tougher audience. Overseas, we are teaching foreign nationals with language barriers usually, a 2.5 on a level of 10 in reference material and student quality. In America, these instructors will now face highly trained law enforcement officers who work the streets every day in a high threat level against sophisticated and violent opponents.

Instructors will have to put professionalism over profit. These trainers will either raise their standards to meet the training needs of these officers or fall by the wayside. They will have to make tough choices and curb their egos to survive. If they are smart, they will look at the current tactics and techniques and improve them. If they do not, they will try and bring in radically different systems that do not apply.

To survive, these instructors will have to put their ego aside and solicit constructive criticism from each class, go back and attempt to make the next one better. Failure to do this will alienate students. The word of mouth in the law enforcement community will kill them. I have had students reporting back daily to their coworkers on the class quality and content. In the civilian arena, I have read students' daily posts on Internet forums while the class was still being conducted.

## HISTORY

I began my training profession as I was nearing military retirement. During the first six months of my new business, I saw the reports of several tactical officers dying due to training accidents. Some of these incidents were captured on video that I still use in my instruction. The first class I remember developing was weapon safety which took the "big four safety rules" and explained them and expanded upon the rules to include CQB and working on downed officers and suspects.

I also began to write articles on weapon safety for tactical officer journals to help educate and address the problem. It has taken nine years, but I finally believe that most of the law enforcement community is on board with these protocols.

When I first started, I wanted to change the world, but knew I could not do so overnight. I started by driving from training site to training site across the country with the equipment I needed in the back of my truck. I would try and coordinate trips to spend two weeks in the area to reach more students and make the trip cost-effective. I drove because I brought too much equipment to ship ahead or for flying. I always thought it was important to bring training aids to ensure that students had a proper example for the day they will become a trainer. Simple items such as dummy flash bangs and flex ties were necessary for students to practice all the actions that they would perform on real operations. Flash bangs deployment must be practiced along with putting hands-on people. I brought the tools to ensure all this took place and showed students that they could make the same props and train with them once I left.

For the first year of teaching a class, I would hand out critique sheets after each class and diligently read them to make the class better. At night in the hotel room, I would review and recover video I wanted to use in classes to make my presentations better. Years later, I traveled to a local agency in my state and offered to train them for a day in exchange for the opportunity to choreograph a few drills for my PowerPoint classes. I wanted students to see the perfect, or almost perfect, way to execute a drill or maneuver (while in the classroom environment).

In the end, I was able to refine my training modules and instruction to build over twenty-two classes. Many of the training modules fit into the other blocks of instruction and complement one another.

## ETHICS

Staying on the path and not falling into the ditch from time to time was my goal. I used the "rucksack" analogy in my first book to illustrate my journey. That being, I see my place in this profession as a slow climb up a hill. My rucksack is filled with positive attributes and values. As the hill becomes steeper, I don't want to lighten my load of these qualities to make it to the top only to find an empty rucksack.

Some instructors come into this arena with the objective to make a profit grossly outweighing the desire to teach and give back. You, the student, will sort them out quickly and will not call them back for follow-on training. These "one hit wonders" will seal their own fate in regard to training and work ethic.

## SAFETY

I can determine the competency and professionalism of an agency's officers in a few minutes by watching their weapon-handling skills. It is critical that instructors have and promote positive and aggressive safety protocols in training both during classroom, flat range, and tactical training. Students, you should ask potential instructors about their safety protocols or regime. If they do not have one, be suspicious.

Teaching safety in weapon handling and training exercises while conducting tactical operations is critical to survival of the personnel, department, and training programs. Ensure these standards are maintained and enforced.

## KEY POINTS

- Put professionalism before profit.
- Maintain ethical standards.
- Ensure safety is adhered to, refined, and maintained.

# GLOSSARY OF TERMS AND ABBREVIATIONS

| | |
|---|---|
| **"G"** | Guerrilla fighter |
| **AAR** | After-Action Review |
| **AC** | Assault Commander |
| **AD** | Accidental Discharge |
| **AFC** | Assault Force Commander |
| **AIT** | Advanced Individual Training |
| **ATL** | Assistant Team Leader |
| **C & S** | Command and control |
| **CCT** | Combat Control Team-Air Force |
| **concealment** | A visual barrier that will hide an individual, but which cannot withstand bullets or other projectiles |
| **cover** | A manmade or natural object that will stop bullets, shrapnel, or other projectiles |
| **CP** | Command Post |
| **CQB** | Close Quarters Combat |
| **CSAT** | Combat Shooting and Tactics |
| **DOD** | Department of Defense |
| **DOS** | Department of State |
| **dry-practice/ fire** | Training with an unloaded weapon that helps focus on muscle memory |
| **EMT** | Emergency Medical Technician |

| | |
|---|---|
| **Ex-fil** | Slang for exfiltration |
| **FTO** | Field Training Officer |
| **HALO** | High Altitude Low Opening |
| **HMMWV** | High Mobility Multipurpose Wheeled Vehicle |
| **IAD** | Immediate Action Drill |
| **IED** | Improvised Explosive Device |
| **LAW** | Light Anti-tank Weapon |
| **LCE/LBE** | Load Carrying Equipment/Load Bearing Equipment |
| **NCO** | Non-Commissioned Officer |
| **NFDD** | Noise Flash Diversionary Devices |
| **NVG** | Night Vision Goggles |
| **PJ** | Pararescue Jumper Air Force |
| **ROE** | Rules of Engagement |
| **ROTC** | Reserve Officer Training Corps |
| **RPG** | Rocket Propelled Grenade |
| **Ruck** | Common slang for "rucksack" |
| **SAW** | Squad Automatic Weapon |
| **sector** | Assigned area |
| **Simunition** | Specially designed munitions which allow students to participate in realistic training scenarios and drills. |
| **SJA** | Staff Judge Advocate |
| **SNA** | Somali National Alliance |
| **SOP** | Standard Operating Procedure |
| **TC** | Track Commander |
| **TCP** | Traffic Control Point |
| **TL** | Team Leader |
| **TOC** | Tactical Operations Center |